# IQ, Heritability and Racism

# IQ, HERITABILITY AND RACISM

## A Marxist Critique of Jensenism

by James M. Lawler

with an introduction by Brian Simon

LAWRENCE & WISHART, London

SBN 85315 428 7
© 1978 by International Publishers
Introduction © 1978 by Brian Simon
First Edition, 1978
Printed in the United States of America

# Contents

*Foreword* / ix

*Introduction* / 3

**1 • Jensen's claims** / 7
 *Class and race differences in IQ* / 8
 *What is IQ?* / 11

**2 • General theory and method of IQ** / 15
 *Intelligence: fixed or developing* / 15
 *Sponge theory: innate and fixed capacity* / 17
 *Philosophical implications* / 18
 *"Intelligence is what intelligence tests measure"* / 20
 *A priori method* / 24

**3 • Some IQ questions** / 29
 *Examples of IQ test items* / 29
 *Questions about these items* / 33

**4 • History of IQ theory** / 39
 *Galton, father of eugenics* / 39
 *Alfred Binet* / 42
 *Lewis Terman and the Stanford-Binet*
  *Intelligence Scale* / 46

**5 • "Reliability" and "validity" of IQ tests** / 53
 *Reliability* / 53
 *Validity* / 55
 *Content validity* / 55
 *Predictive validity* / 57
 *Concurrent validity* / 59

*Construct validity / 61*
*Jensen's arguments for validity: "g"*
*and Headstart / 62*
*Behaviorist interpretation / 65*

**6 • Differences: in the children or in the schools? / 69**
*"Fluid" and "crystallized" intelligence / 69*
*"Culture-fair" tests / 71*
*Literate intelligence: a product of history / 74*
*Thinking can be taught / 78*
*Inequality of opportunity / 81*
*IQ, tracking and segregation / 83*

**7 • Dialectical relation of biology and society / 87**
*Darwin and Social Darwinism / 89*
*Biology and human evolution / 91*
*Biological changes in human evolution / 96*
*Race / 97*
*Alienation and racism / 99*

**8 • Relative and absolute differences / 103**
*Blind relativity / 103*
*Basic historical differences / 108*
*Genius and history / 112*
*Population: concrete or abstract / 115*
*Jensen: from relativism to absolutism / 116*
*Difference and identity / 123*
*Equality and inequality / 128*

**9 • Heritability / 133**
*From innate IQ to IQ heritability / 133*
*Heritability and heredity / 135*
*Jensen's "sociology" / 139*
*Heritability and widthability / 142*
*How much or how? / 146*
*Interaction, development and the metaphysics*
*of variation / 148*
*Mendel and evolutionary interaction / 151*
*The social environment and human activity / 156*

**10 • Twins and other relations / 159**
*Is human heritability possible? / 161*
*Twin studies / 163*
*Kamin's analysis of the data / 165*
*Kinship correlations and adopted children / 168*

**11** • ***Real science and real freedom*** / *173*

      *The practical reality of racism* / *176*

      *Too little intelligence or too much?* / *180*

**Notes** / *183*

**Index** / *191*

# *Foreword*

*T*here is, perhaps, no more important question than that of human intellectual capacity, and especially the degree to which it is capable of change and development. Certainly our conceptions of this must affect—even determine—our attitude to education and its power (or otherwise) to promote such development. Hence the significance of the nature-nurture controversy, revived over the last decade by Arthur Jensen in the United States, and by the late Sir Cyril Burt, Professor Eysenck and others in Britain. This group of psychometrists presents what might be called the hard line, 'classic' view which claims that intellectual capacity is very largely determined by genetic endowment.

In Britain we are only now just breaking free from an educational system that found its rationale, or justification, precisely in theories about human capacities derived from mental testing. In the nineteen thirties and forties, when the system was being constructed, psychometrists maintained that intellectual capacity was *wholly* due to genetic endowment and was therefore fixed, unchanging and, in addition, accurately measurable by group intelligence tests. This was the advice tendered to, and accepted by, the Consultative Committee to the Board of Education whose series of reports (Hadow, 1926; *The Primary School,* 1931; Spens, 1938) underlay the structuring of the divided system of secondary education with its concomitant, streaming in the primary school and early selection at 11 plus for secondary education.

For some years from the late nineteen fifties or early sixties what may be called the hereditary thesis was under a cloud, and in fact rejected in government reports concerned with education (Robbins, 1963; Newsom, 1963; Plowden, 1967). At this stage these ideas seemed discredited, as a result both of theoretical criticism of the techniques and assumptions embodied in testing, and of their apparent failure to provide practical guidelines to educational procedures. Much greater emphasis was now put on the importance of education, and of the child's life experiences, in determining intellectual growth. This crucial shift of view underlay the transition to comprehensive education, accepted as national policy in 1965.

However, early in 1969, Arthur Jensen published his now famous article, restating original views, in the *Harvard Educational Review*. In the American context this had relevance, particularly, to theories concerning racial differences in 'Intelligence'. Jensen's views were propagated in Britain more specifically in the social class/educational context, as exemplified, in particular, by the contributions of Burt and others to the series of *Black Papers* on education from 1970 on. Today the fascist press in Britain makes considerable use of these restated theories in an overtly racist sense.

Intelligence testing is, therefore, once more, an important theoretical question in Britain, and one having serious implications in terms of social policy. For this reason Professor Lawler's scholarly critique is to be warmly welcomed. As a philosopher, Professor Lawler brings a sharply analytic mind to bear on some of the key questions underlying the theory and practice of testing. What precisely is meant by the term 'Intelligence'? How is an Intelligence test constructed and what are the hidden assumptions embodied in its technology? One feature of this book is the width of its range; Professor Lawler is familiar with the history of mental testing, with its relation to Social Darwinism and racial theories, as well as with the technical literature of psychometry, in particular the supposedly crucial twin studies of Burt and others. His chapter on heritability is especially notable for its clarity of exposition of a complex issue.

Professor Lawler approaches his critique as a Marxist. The strength and subtlety of his analysis bears witness to the power of this approach when applied to an area that cuts across normal disciplinary boundaries, having both theoretical and practical implications. His book will repay the most careful and detailed study by those specifically concerned with education and psychology (and their interface). It should also help to clear up contemporary misconceptions as to how Marxists approach the whole complex question of human intellectual capacities.

BRIAN SIMON
*Professor of Education*
*University of Leicester*

*January 1978*

# IQ, Heritability and Racism

# Introduction

*T*heories of the innate intellectual inferiority of working class and especially of Black people attack the central core of the work of our school system. Where educational institutions aim at the *development* of intellectual and cultural abilities, the main thrust of the theories of Arthur Jensen, Richard Herrnstein, Hans J. Eysenck, William Shockley and others, is that the level of intelligence that students will reach or fail to reach was basically decided once and for all in the genes. Schools should therefore not be thought of as providing an enriching and creative environment, but should be adjusted to the function of sorting out and selecting the "bright" from the "dull," as determined by nature, and as basically reflected in the existing social hierarchy.

For teachers in the schools attended by working-class and Black children a fatalistic attitude is cultivated by supposed experts in matters of intelligence. According to some, the difficulties encountered in teaching cannot be resolved by such "superficial" methods as improving teacher-student ratios, or providing early childhood education, or improving pedagogical understanding of the dynamics of learning. According to Jensen, that which is essential to the educational process, thinking itself, cannot be taught, because it is a capacity that is fixed and determined by biological mechanisms.

For the children who fail to move ahead normally in their studies, spokesmen for the newest "science" essentially tell them that they face nearly insurmountable obstacles because they were "born dumb," because they "don't have the brains." And for the

*3*

parents who had hoped that their children would acquire the intellectual means of improving their lives, who make sacrifices and are willing to sacrifice more for the future of their children, the media announces that the real "brains" in this world think that they are wasting their time, and might as well face the fact that they have only endowed their offspring with a deficient amount of intelligence genes.

The 1954 Supreme Court decision condemning the legal apartheid educational system in the U.S. South has led to the growing recognition that segregated schools in the North and elsewhere are no less the result of racist planning. Segregated schools are inherently unequal, and, while perpetuating inferior education for Black children, also condemn white children to the backward cultural isolation in which poisonous racist ideas easily breed. The theory perpetuated by Jensen, however, locates inequality and educational inferiority in the genes of the victims of racism themselves, and declares nature itself to be the first cause and champion of segregation.

As the economic crisis facing our country deepens, there is a message for all who are concerned with maintaining and extending the school system. In the competition for funds, education is said to be more of a luxury than a necessity. Early childhood education and early compensatory programs, as well as integration efforts, won't change the picture much, it is argued, so the money for these programs is better spent elsewhere. While hundreds of billions of dollars are poured into the military-industrial complex, projects for further expanding the scope of higher education, for establishing general policies of open admissions and free college and university programs, are declared utopian and unproductive. Are the doors of the campuses closing on more and more young people who cannot afford the high cost of living and studying at the same time? This is unfortunate from the point of view of "ideals," sigh the theorists of the stagnation and recession of intelligence. But perhaps it's for the best, they argue, since if a person doesn't have the money, he probably doesn't have the brains either.

The impact of the theories that will be discussed in this book is not limited to the school system, but has a much broader effect. The main thrust, as has already been pointed out by many critics, is a reinforcement of racism with the prestige of science. The highly technical arguments that Jensen marshals forth to establish the "plausibility" of his thesis may go over the heads of academics, as well as of students and the lay person who read about it in *Time* or *Atlantic Monthly*. But the conclusion Jensen draws filters down just as definitely into the ordinary pollution of racism.

On the other hand, 1) if intelligence is not innate, but a social, historical product, 2) if a truly scientific pedagogy locates the obstacles to learning in the practical environments of the children, both in and out of school, and demonstrates that children are all capable of assimilating the essentials of a scientific culture, 3) if this is not the best of all worlds but one in which the talents of people are systematically wasted and destroyed, and 4) if the potentialities of society today make possible and positively require a broader and deeper cultivation of a scientific culture among the entire population—if these things are true, then what follows is not resignation, passivity, fatalism and despair, but recognition of real possibilities, an active approach to education, outrage at the injustices committed, and determination to fight for the educational rights of the people and the children.

These are not purely theoretical matters; they have the kind of practical import which Marx had in mind when he wrote that "theory becomes a material force when it is grasped by the masses."

This is not to deny the special character of the theoretical aspects of the issues raised by theories of IQ and its "heritability." There are many questions that come to mind when we begin to examine beliefs that have been so deeply ingrained in us as to be part of colloquial language. There are real problems still to be mastered in the complex realm of the theory and practice of education. It is necessary to understand the relation of biological and social factors of human development; the causes and consequences of the formation of races; the historical growth of human thought; the division of intellectual and manual labor; the causes and effects of racism; the laws of the development of cognition in the child and the practical application of these laws in scientific pedagogy; the structure of the school system in the U.S. today; government policy and priorities in relation to education; and the demands and possibilities of modern science and technology. These are all matters that require understanding, and call on the resources of many different disciplines.

As to the concepts most closely related to Jensen and his followers, the issues involved in IQ measurement and heritability estimates are extremely confusing for both experts and non-experts alike. Lack of understanding of historical materialism and dialectical theory of scientific methodology produces a theoretical incapacity when interpreting the meaning of certain empirical facts and methods of analysis. Jensen belittles "philosophizing" and "ideological" approaches to the study of empirical data. But there are no purely empirical approaches to reality which do not make use of certain categories and general methods of analysis. Unless these categories and methods are critically examined, they are inevitably borrowed from uncritical philosophy and are subject to the effects of prevailing ideology, whether consciously or not.

The object of this essay is to examine the basic concepts and methods that are most pertinent to the arguments which Jensen and others have put forward, to expose their philosophical and ideological presuppositions and prejudices, and to outline an alternative interpretation based on the concepts of dialectical and historical materialism. This Marxist critique does not pretend to substitute for empirical research but to facilitate such research by removing the theoretical blinders that operate in much of the literature on IQ and its causes.

# 1

## *Jensen's claims*

*In* 1969 Arthur Jensen wrote an article in the *Harvard Educational Review* entitled "How Much Can We Boost IQ and Scholastic Achievement?" This article began with the assertion that "Compensatory education has been tried and apparently it has failed."[1] What it has failed to do is to change the scores on IQ tests of "underprivileged children," particularly Black children. As measured by IQ tests Black children score an average of fifteen points below white children. "Environmentalists" attribute these differences to social and economic disadvantages, leading to poorer performances in school and on IQ tests. According to Jensen, such a theory justified massive compensatory educational programs designed to equalize opportunities, and bring up the average performance of the underprivileged, especially Black children. The failure of these programs to produce long range improvements as measured by IQ tests, Jensen argues, must lead us to question the presuppositions on which such programs are based, particularly the argument that IQ differences between races are due to environmental differences. Perhaps, Jensen argues, the differences are innate in the children themselves.

This "hypothesis" leads Jensen to a lengthy discussion centered around two major areas: 1) IQ tests as valid measurements of intelligence; and 2) methods of estimating the "heritability" of IQ. IQ "heritability" is the extent to which IQ differences are due to genetic differences; Jensen argues that IQ has a heritability of

7

80%—only 20% of IQ differences being due to differences in the environment. He concludes with academic circumlocutions that IQ differences between Blacks and whites are "probably" due to genetic causes.

## Class and race differences in IQ

The starting point of Jensen's argument is that there is a consistent finding of racial and class differences on IQ tests. Jensen focuses primarily on the differences between the average scores of Blacks and whites. Herrnstein, in *IQ and the Meritocracy,* stresses class differences. Upper and middle class children score significantly higher than working class children. Jensen accepts this class difference, too. Both writers attempt to explain these differences in IQ scores on biological, rather than sociological or "environmental" grounds. In the relation between "nature" and "nurture" or between "heredity" and "environment," the Jensenites argue clearly for the primacy of biology or heredity in explaining differences in "intelligence" as measured by IQ tests.

The fact that the theories which are examined in this book argue for the biological inferiority of the intelligence of races *and* classes is very significant. While the primary focus of attention in the popular media has been on the racist aspect of the question, it is important to recognize that it is also a class question. While the racist edge is foremost, the theory and its applications cut into the educational and social prospects of the majority of working people.

Thus, according to Jensen studies have shown that Blacks on the average score 15 IQ points lower than whites, while the difference between socio-economic classes range from 15 to 30 points.[2] Eysenck, whom Jensen credits with the best popular presentation of his ideas,[3] cites tables from the English IQ theorist Sir Cyril Burt that show a difference of 55 IQ points between upper level professionals and unskilled workers.[4]

One might think that these differences can easily be explained by the circumstances in which individuals in different classes have been raised, and by the particular effects which racial discrimination adds to the largely working class Black population. This point of view, that the economic, social and cultural "environment" explains these differences, is held by a large majority of social scientists as well. So much is this so that Jensen has been hailed as a pioneer and an intellectual hero who has challenged the prevailing "orthodoxy."[5] Scathing criticism is leveled against "dogmatic environmentalists" who will not examine the latest findings of modern science. According to Jensen "social scientists and educators have been indoctrinated to ignore genetics, or to believe that genetic factors are of little or no importance in human behavior and human differences . . . ." He calls for a "revolution in our thinking," comparable to the scientific revolutions of Copernicus, Darwin and Einstein.[6]

Richard Herrnstein begins his book on the trend toward a genetically determined caste system with a "True Tale From the Annals of Orthodoxy." There he describes his own "persecution" not only by standard environmentalists, but also by "Marxist visionaries": "It is hard to argue that the 'class struggle' can be resolved by a redistribution of wealth and capital, if it should turn out that something more than economics distinguishes the contending classes."[7] Referring to the banning of IQ tests in the U.S.S.R. in 1936, Eysenck argues that test results had run counter to the "doctrine of Communist egalitarianism‹ and the myth of omnipotent 'Soviet man.'"[8]

Theorists of the French and American revolutions suffer no less than the theorists of the socialist revolutions. In *The Inequality of Man,* Eysenck takes issue with the egalitarianism of Rousseau and Jefferson. In his *Atlantic Monthly* article, "IQ," Herrnstein launches an attack of the American Declaration of Independence as well as the *Communist Manifesto*:

> The specter of Communism was haunting Europe, said Karl Marx and Frederick Engels in 1848. They could point to the rise of egalitarianism for proof. From Jefferson's "self-evident truth" of man's equality, to France's *"égalite"* and beyond that to the revolutions that swept Europe as Marx and Engels were proclaiming their *Manifesto,* the central political fact of their times, and ours, has been the rejection of aristocracies and privileged classes, of special rights for "special" people. The vision of a classless society was the keystone of the Declaration of Independence as well as the Communist Manifesto, however different the plan for achieving it.

Against this background, the main significance of intelligence testing is what it says about a society built around human inequalities.[9]

The view that substantial IQ differences can be explained by "environment" is shared by many social scientists, Marxist and non-Marxist. We will see later that there are particular weaknesses in non-Marxist forms of environmentalism which make them vulnerable to some of the criticisms of environmentalism which are made in defense of the biological, genetic explanation of IQ. At the same time, there is no doubt that partly as a result of the battle against racist and fascist eugenic theory, many Marxists, especially those under the influence of Lysenkoism, have been guilty of a one-sided view of the nature of genetics.[10] Nevertheless the main thrust of the Marxist rejection of the "genetic explanation" of human evolution and class struggle is more valid than ever. The fact that Herrnstein begins his argument for the biological basis of class society by a denunciation of Marxist theory and the socialist alternative, is testimony to the force these are exerting at this critical juncture in U.S. and world history. This is all the more reason to avoid distorted representations of Marxism, whether by Herrnstein or by some of his "radical" critics.

It should be noted that the U.S.S.R. is not alone in banning IQ tests (though not diagnostic tests with a fundamentally different character). Group IQ tests have been banned in the New York City schools, and the use of IQ and other forms of tests used for tracking, selecting or screening individuals have recently come under strong attack for a variety of reasons and have been subject to various court decisions limiting their use. Recent books on IQ tests describe the change, *From Awe to Anathema*[11] and *The Rise and Partial Fall of the IQ Test*.[12] As indicated in the latter chapter heading, the IQ test, and particularly the concepts and implications that have traditionally surrounded the IQ test, still maintain a certain vigor. The mystique of IQ seems still to be maintained in at least one compartment of academic as well as popular thought, even as social and intellectual developments are undermining it. Nevertheless, in view of the considerable opposition it is possible that the theories we are now examining constitute the last real stand of the notion of IQ tests as valid measurements of intelligence, defined as an innate, basically fixed capacity for abstract thought.

## What is IQ?

The fact of IQ differences between races and social classes leads Jensen to examine the reasons for the differences. Two opposed options are clear: either inequality is basically socially derived and can be overcome by social reorganization, or it is biologically derived—and society should adjust its educational practices and ideological beliefs to this fact, stressing biological methods of change (eugenics).

The main thrust of Jensen's work consists in arguing for the "genetic explanation." Jensen argues that differences in IQ are 80% "heritable." There are many ways of estimating "heritability," one of the most notable being the study of the IQs of identical twins raised in separate environments. We will examine the meaning of the concept of "heritability" as well as the "twin studies" and other methods of investigating the extent to which genetic differences play a role in IQ. On a more general plane, we will examine the relation between biology and the development of human society. But first it will be necessary to look at the concept of IQ. As we have noted, IQ tests have come increasingly under fire. The notion of a fixed or immutable IQ has been challenged from many quarters. IQ theory itself is on the defensive. Before the cause of IQ differences .can be examined it is necessary for defenders of its genetic causality to explain the significance of IQ tests in the first place.

Thus the fact of IQ differences cannot really be accepted as a starting point. It is necessary to examine the concepts and the methods employed in getting these results. Here we will find that there is already an implicit theory and philosophy that guides the setting up of the tests themselves. The supposed conclusion, that IQ measures an innate, fixed capacity, is basically assumed in the way IQ is set up in the first place.

Jensen and his followers intimidate their critics by referring to specialized statistical methods which supposedly provide objective and scientific proofs of their theses. Thus, arguing that it has been scientifically demonstrated that differences in intelligence "are largely due to hereditary causes," H. J. Eysenck describes those who challenge this assertion as being ignorant "alike of psychometric techniques of intelligence testing and biometric techniques of genetic analysis," and finds it unreasonable to "discuss the problem or write about it, when one cannot tell the difference between epistasis and meiotosis, reduce a Hessenberg matrix, or determine an Eigenvalue."[13] And yet it is the basic presuppositions of psychometric analysis (intelligence testing), rather than the technical features of the statistical analysis, that most often go unquestioned. When venturing out onto more flimsy philosophical grounds, Jensen and his followers leave the safer shores of statistical method and find themselves obliged to explain what these methods imply, their scope and limitations. Here the layman deserves an explanation. Jensen in fact refers to Eysenck as one who has provided a presentation that avoids "the technical yet being accurate, and could well serve as a non-technical introduction to the present work . . . "[14]

To deal clearly with the issues raised by Jensen, and with the theory and practice of IQ in general, it is necessary to have a clear idea of what an IQ test in fact is. Instead of making a fetish of statistical techniques, it is necessary to know what it is that is being measured, as well as the implications of the methods of measurement themselves. One critic describes the "preoccupation with the techniques of measurement" of the psychometrists, which "has meant that relatively little attention has been paid to the substantial content of the tests, and to their validity as tests of intelligence." Far from being impressed with the scientific character of IQ theory, she notes "the confusion about defining intelligence and also the somewhat *ad hoc* and atheoretical criteria used in the selection of items for inclusion in tests." In general, there is "much mystification about what IQ scores mean, mystification which is unfortunately increased by the internal technicalities of quantification that give intelligence tests a largely spurious appearance of scientific respectability."[15]

Critics of Jensen who do not have the technical expertise demanded above by Eysenck may find some comfort in the fact that Jensenism involves broad generalizations overlapping many areas of thought, from general epistemology through biology and physiology, into anthropology, history, sociology, economics as well as psychology, down to educational policy and practical politics. In this short book we hope to provide an introductory approach to the many-sided question of human intelligence, using basic concepts of

Marxism that have been developed in the past and are still developing. At the same time certain questions of technical method must be examined in a way which we hope will be clear and comprehensive.

# 2

## *General theory and method of IQ*

*T*here are few notions so wrapped in mystery and awe as is the notion of IQ. Who cannot remember having experienced a certain shudder of apprehension on first coming to understand that he or she has an IQ which may be secretly buried in the files of the teacher or principal? Is my IQ high, low, or is it "just average"? Most school children have to reconcile themselves to this question, and in many cases to its answer. In general the notion evokes a sense of fatality that is similar to the Calvinistic idea that some of us are predestined to succeed both in this life and the next, while others are born to more modest fortunes and even outright damnation. While this idea of religion fails to find many followers today, its counterpart, cloaked in the prestige of science, is deeply rooted in common belief.

### *Intelligence: fixed or developing?*

According to this common belief, each of us is born with a certain amount of intelligence. IQ is a measure of the amount of that intelligence. A low IQ means that we were born with a low amount of intelligence, while a high IQ means that we were born with a high level of intelligence. Ordinary speech contains many expressions that imply the idea of our having a certain amount of intelligence that is fixed at birth. Someone is a "brain." Another has a "good head for math." Someone else is "gifted" for music. On the other hand, there are the "bird brains" and the "dumb-heads." These

*15*

expressions all imply a notion that abilities or talents are innate, basically biological and fixed. IQ appears to be a technical, "scientific-sounding" expression that implies the same thing. In fact the idea of IQ appears to dignify a popular belief with some degree of scientific status.

Against the notion that we are born with a certain amount of intelligence, there are other notions that stand in opposition to this. We think of science as involving a progressive development of thought, and by acquiring the truths of modern science we think of ourselves as knowing more about life than our ancestors did. Is this not a growth of intelligence? Not only do we think of scientific knowledge as something that grows, we think of our own lives as involving a considerable degree of growth in intelligence. We have learned, whether from experience or from study, and we attempt to overcome mistakes that were made because of earlier ignorance or stupidity.

Two opposed notions of intelligence seem to compete for our adherence: one that involves an inborn, fixed ability, and the other that implies development and learning from life and from science. It is the second notion that appears to be on the side of education and science, whereas the first, with its implications of helplessness and fate, seems to be more closely related to certain forms of religious beliefs. And yet the supposedly scientific idea of IQ is generally understood to uphold the idea that intelligence is basically something that we were born with and which does not change throughout our life. And it is precisely in the field of education, where we might have expected to find a preference for a developmental notion of intelligence, that IQ tests and scores play the greatest role.

So, we may be tempted to think that as we passed from one primary school grade to another, and from primary to secondary school, we were continually developing our intelligence. We may think that higher education opens up even broader horizons for the development of our minds, and that learning can develop throughout life. If we are tempted to think this way, we are reminded by our fixed IQ that our "intelligence" has not budged one iota throughout this entire process. For if the IQ test has any validity, we will have about the same IQ score at the age of five, at the beginning of the school age, as at the age of eighteen or twenty-five, at the end.

## Sponge theory: innate and fixed capacity

What has taken place during this process, if our "intelligence" has remained the same despite all our efforts? As we confront these contrary ideas of intelligence we may think of intelligence as an inborn *"aptitude"* to learn which is different for different people. The person with the high IQ can learn more and go farther than the one with the low IQ. The brains of different people have different limits as to how much they can absorb. Like sponges, some brains can soak up more knowledge than others. According to this theory, IQ not only stays the same throughout the educational process, it also draws the limit as to how far we can go. Thus, Eysenck writes: "What the figures show is simply that absence of a given level of intellectual ability (low IQ) is an almost absolute bar to progress towards higher education (as well as being usually coupled with a dislike of further education, and a strong desire to get out of school and earn a living!)"[1]

The sponge theory of intelligence says that our minds can "develop," but only to a certain point, and that point is fixed at conception. There is a "given level of intellectual ability" which prevents some from going as far as others, and is even coupled, according to Eysenck, with a desire to leave school when the sponge is filled, and get a job.

The theory that IQ is a kind of inborn "ceiling" to our possible development seems to agree with certain facts and feelings. It is true that some people seem to grasp intellectual subjects with ease, while others struggle, sometimes in vain, to master the same material. Perhaps we have said to ourselves at one time or another that we "just don't have it." And so many young people, faced with frustration in school work, may in fact feel that they would do better getting a job and making some money. (On the other hand, pressures to make money may distract many others from giving their full attention to school. Also the increasingly higher costs of continuing education, coupled with diminishing job opportunities, frustrate the desire to learn.) The fact that some students easily master subjects while others fail no matter how hard they try may lead many to conclude that some have what it takes and others don't. The experience of failure, without the understanding of *why* it has occurred, reinforces the idea that there is an inborn deficiency in intellectual aptitude in some people and an innate superiority in others.

There are other facts, however, which point in the opposite direction. When we look at the fact that the average education for the whole population has risen dramatically in the past fifty years we may question whether people's brains have increased in capacity. Who could have predicted, in the depression years of the thirties, what a tremendous boom would take place in the field of education and the large numbers of children who would attend college from families in which the parents might not have completed high school? There was a time not too long ago when few people were thought to "have what it takes" to finish high school. Such a sentiment reflected experiences similar to those described above— the sentiment of failure and frustration in school, combined with the need, if not the desire, to work at very early ages. If "common sense" was wrong in the thirties, perhaps it is wrong today when cutbacks in all levels of education, in conditions of major economic crisis, are spawning a renewal of the biological and sociobiological "explanations" of the difficulties young people and the schools are facing. If under different conditions, larger numbers of young people were able to attend college, perhaps under other conditions the vast majority may "have what it takes" to complete a higher education.

### Philosophical implications

On the philosophical level these two trends are reflected, in the first place, in the opposition between the "dialectical" and the "metaphysical" view of reality and of its reflection in people's minds—i.e., intelligence. This opposition was stated clearly and in a straightforward manner by Frederick Engels who described "The old method of investigation and thought which Hegel calls 'metaphysical,' which preferred to investigate *things* as given, as fixed and stable, a method the relics of which still strongly haunt people's minds. . . ."[2] The IQ view of intelligence is precisely that of something "given, fixed and stable," something that contrasts with the dialectical view of the world which Engels describes as "The great basic thought that the world is not to be comprehended as a complex of ready-made *things,* but as a complex of processes, in which the things apparently stable no less than their mind-images in our heads, the concepts, go through an uninterrupted change of coming into being and passing away, in which in spite of all seeming accidents and of all temporary retrogression, a progressive development asserts itself in the end . . . ."[3]

The impression described above that there are two opposite trends in people's general concepts about intelligence is expressed on a philosophical plane in this contrast between "metaphysical" and "dialectical" views of intelligence. When this opposition is described in such general terms, it appears to be almost a matter of common sense that the dialectical or developmental view corresponds most adequately with reality. Yet it is amazing that a concept as metaphysical as that of IQ—of intelligence as something fixed and stable—should have such a strong hold on people's minds and should be defended still in the strongholds of certain sectors of psychology that deal with the measurement of intelligence.

A second contrast of philosophical orientations overlaps with this first. The IQ concept of intelligence is not only metaphysical, in that it describes intelligence as something fixed and stable. "Intelligence" as defined by Jensen is not a "mind-image" of reality, but an inner "capacity" with a definite "ceiling" for possible development and learning. It is not something that *reflects* reality or "environment," but an inner capacity that precedes and to a great extent determines the social environment of the individual. In philosophical terms Jensen's metaphysical concept of intelligence is basically "idealistic," rather than "materialistic."

Although a certain materialistic aura seems to surround Jensen's biological definition of intelligence, philosophical materialism has two sides. Engels expresses the opposition between idealism and materialism in philosophy in terms which again are clear and straightforward. On the one hand there is the question of the relation of nature and spirit (consciousness, intelligence). Those who assert the primacy of spirit to nature fall into the camp of idealists. Those who assert the primacy of nature and for whom "consciousness and thinking, however supra-sensuous they may seem, are the product of a material, bodily organ, the brain" fall into the camp of materialists.[4] Jensen stresses the materialistic basis of thought in the brain, but in a "metaphysical," static and reductionist manner, as will be discussed in detail later. The other side of materialism is expressed in the relation between thought and reality, subject and object. Those who assert that our minds basically determine reality, or that ideas and categories preexisting in our minds determine our knowledge of being are philosophical (epistemological) idealists, while those who assert that intelligence is primarily a reflection of reality, that we are capable of knowing objects or processes that have their own existence and laws, are "materialists." On this side of the question, IQ theory of minds as having different innate capacities, which predetermine from inside the head of the person how much or how well one shall know, is clearly idealistic. Ultimately it is this clearly idealistic or subjectivistic side of the question that predominates in the IQ ideology. In measuring intelligence, it is not *what* a person knows that counts, but an indefinable inner "something" that predetermines the form, scope and range of thought. Jensenism is a theory of "innate ideas" which have been located, not in a detached "spirit" as in the classical philosophical version, but in the biological basis of thought itself.

These brief definitions are meant to formulate somewhat more precisely the theoretical concepts involved in the conflicting trends described at first in an impressionistic manner. Jensen is severe with philosophers who speculate about the relation between intelligence and its biological and social factors without looking at the facts. We do not wish here to settle matters with philosophical definitions. General theory is valid only when it explains the facts and is verified in practice. On the other hand, we do not believe that it is possible to explain the facts and to verify hypotheses without resting on some general theory. By bringing out the opposite tendencies of philosophical thought in a simple and general manner we hope to provide a frame of reference which will enable us to interpret the facts, and the method employed in arranging them, with full critical awareness of the concepts and methods we employ. We will see that with respect to IQ theory and heritability measurements Jensen and his followers offer anything but such critical clarity.

## *"Intelligence is what intelligence tests measure"*

In the popular view of it, IQ is a measurement of the amount of intelligence that a person has. As to what intelligence itself is there may be a vague concept of a capacity to solve problems of a theoretical sort, just quick-wittedness and being "bright," or knowledgeability in general. There is a certain "common sense" view of intelligence which may, upon analysis, be complex and contradictory and closely related to certain practical situations. An "intelligent" person in one situation exhibits certain characteristics relating to that situation. There are intelligent athletes, intelligent trade unionists and intelligent bankers. There are also, of course, intelligent students. What all of these have in common is their ability to master their particular environment and the problems that are confronted in it. But it is not obvious that their "intelligence" is similar in any other respect—that it is the *same* ability that operates in all cases. We can ask, then, what is the "intelligence" that is measured by a person's IQ?

When we turn to IQ theory, however, we find that this question is answered in a very peculiar way. According to Jensen not only are there many disagreements and arguments among psychologists as to what intelligence is, but "There is no point in arguing the question to which there is no answer, the question of what intelligence *really* is."[5] Instead, he writes, we can take an "operational" stance and define intelligence by the way we measure it, whatever it is. In fact, "intelligence, by definition, is what intelligence tests measure."[6] The only thing we know for certain about intelligence is the fact that we can measure it. In this response, we seem to have come full circle. We have asked what intelligence is, so that we might know how to measure it, and have been told that intelligence is defined by our manner of measuring it.

This approach to intelligence clearly involves a version of philosophical idealism or subjectivism. In the relationship between ideas and the reality they reflect, Jensen asserts the primacy of the ideas.

Eysenck puts this question squarely: "We are being asked, in effect, how we know that our measurements stack up against something existing outside these measurements. This, of course, is an impossible question.... Intelligence is not a 'thing,' but a concept—just as gravitation is a concept, or heat."[7] He continues to argue that in the history of science many theories were developed "operationally"—i.e., in relation to methods of measuring a phenomenon. Electricity was first understood in relation to its effects in heating a wire or in moving the needle of a compass. There were methods of measuring heat long before there was a deeper scientific knowledge about the nature of heat. He admits that "Many years after Faraday's great work, and Maxwell's theoretical and mathematical labors, we have a theory [of electricity] which we can offer the inquirer, and which he may find more satisfying," but adds that prior to this point "it is doubtful if Faraday, Ampere, and Ohm would have offered him anything beyond a statement that electricity was what electricity tests measure."[8]

The assertion that intelligence, like heat, is a concept, not a thing, reflects a fundamental ambiguity or a kind of duplicity and slipperiness. There is, it would seem, something "out there" that *is* heat. And then there is our *concept* of heat. Of course we cannot *know* anything about heat without some intuition or concept of what heat is and does. Philosophical materialism rests on the fact that there are real processes in the world, and that our ideas, as "mental images," tend to approximate to an understanding of these processes. Moreover, our knowledge of heat is a developing knowledge, rooted in sensation and practical experience, and passing through a long complex history involving mythical, theological and eventually scientific, as well as pragmatically oriented concepts of what heat is. Early man learned how to make fire without a scientific understanding of combustion. Had there been no pragmatic or "operational" understanding of fire—"defined" by the methods used to make it and its practical effects—there would undoubtedly never have arisen a scientific theory of the combustion process The difference between the "operational" understanding and the scientific understanding is one of a *progressive* movement to a more adequate, more essential level of understanding, which grasps phenomena not only in relation to how we use them or in terms of their practical effects, but in terms of their basic laws of development. In this process of the advance and deepening of thought it is necessary for there to be some "thing" or process which our concepts can grasp in a progressively more adequate manner.

Thus a fully understood materialist theory is a dialectical theory of knowledge. Because really existing processes, including mental processes, are acknowledged, it is possible to improve on our concepts of these processes. Knowledge reflects reality in a developing manner, passing through various stages. This is the essence of the dialectical materialist theory of knowledge. The Marxist theory of knowledge has to be distinguished from "contemplative material-

ist" theories, according to which the mind only passively registers the "facts" or "mechanical laws" but is not active in this process and is not capable of facilitating the active transformation of the reality it attempts to understand. In line with general scientific practice, dialectical materialist theory is intimately connected with practice.[9] By having a better understanding of the tendencies, processes and laws of life, as well as of thought, we are in a better position to influence the outcome of these processes. So knowledge of the laws of the development of human intellectual processes, which are intimately connected with social life as a whole, provides a scientific basis for more advanced forms of education.

There is another ambiguity in Eysenck's analysis. Is IQ only at a pre-theoretical stage—so that we should expect some day to find a more adequate theory which would define intelligence in terms of its causes and laws? Neither Jensen nor Eysenck is so modest. It is untrue, Jensen writes, that "intelligence exists only 'by definition' or is merely an insubstantial figment of psychological theory and test construction. Intelligence fully meets the usual scientific criteria for being regarded as an aspect of objective reality, just as much as do atoms, genes, and electromagnetic fields."[10] And Eysenck writes that "even operational definitions incorporate a theory, even though it may not be clearly expressed."[11] He goes on to say that various tests "appear to be measuring the same thing," and concludes that "The empirical data strongly support, and nowhere refute, the notion that our problem-solving behavior in a great variety of different situations can be accounted for, as far as individual differences are concerned, by reference to a concept of general intelligence which is reasonably well measured by traditional intelligence tests."[12] (Is it the *concept* that is measured by intelligence tests, or is it "general intelligence" itself?)

From these passages it appears that for Jensen and Eysenck (and Herrnstein too, as we will see in a different context), the "operational" definition is not considered to be a stage in the development of intelligence theory, but an adequate *basis* for establishing a theory. The "operational definition" will not be replaced by a theoretical definition which will permit a more adequate, scientific grasp of the essential causes and laws of the phenomenon. Such a new scientific understanding would be the basis of new methods of measuring the reality as well as of new forms of practical control of

it. But for Jensen and Eysenck it is the operational method itself, not the reality of intelligence, that provides the basis for developing the "scientific" theory of intelligence. In the analogy with the theory of electricity, Eysenck, in effect, argues that the process of thought consists of 1) the measurement of electricity by the movement of a compass needle when passed through an electromagnetic field, 2) the "definition" of electricity by compass needle movement, and 3) the elaboration of a "theory" of electricity by a generalization from the characteristics of the movement of the compass needle. Such a theory would include an aspect of electrical phenomena—while leaving out other aspects and failing to formulate basic laws for the movement of electricity itself. It would also inextricably involve in its "theory" properties of the compass which are external to electricity.

## A priori method

The method employed here is described by Engels as a form of "the old favorite ideological method, also known as the *a priori* method, which consists in arriving at properties of an object deductively, from the concept of the object, instead of learning them from the object itself. First the concept of the object is formed from the object; then the spit is turned round, and the object is measured by its image, the concept of it. The object is then made to conform to the concept, not the concept to the object."[13]

The ambiguity or duplicity in the use of the terms concept and objectivity are a necessary result of the *a priori,* ideological and subjectivist approach taken by Eysenck and Jensen. IQ tests are a *form* of reflection of intelligence—they are a subjective method of grasping an actually existing phenomenon, the mental capacity or intelligence of people. IQ is or implies a certain concept of intelligence. But instead of measuring the concept by the reality, the spit is turned around and real intelligence is made to fit the concept of it and the method first devised to grasp it. This is the idealist, *a priori* side of the method. The concept becomes the basis for measuring the reality—as is admitted when "intelligence" is "defined" by the method of measuring it. Such a method stifles the development of the science of human thought because it takes one form of approximation to the understanding of intelligence and stops there. This is

the metaphysical, anti-dialectical aspect of IQ method. Any appearance of "development" in IQ theory consists in speculating about the properties of intelligence from the features of the limited form of appraising it. There is thus no movement from an unsatisfactory, limited, but historically necessary phase in the history of the science, to its replacement by a more adequate grasp of the object. Such a movement is only possible if there is some *reality* that exists outside of the method, form of measurement or primitive, "operational" concept.

The question asked by Eysenck is really a necessary one for the progress of any science. There must be some thing or process against which our concepts and methods of measurement must "stack up." Only then can one method improve on another, one conceptual framework make way for a more adequate one, i.e., for concepts that grasp reality, even mental reality, more fundamentally and completely. The IQ test may consist of a first approximation to an understanding of human intelligence. However, because of the "success" of IQ tests, the study of the properties of these tests has replaced the study of intelligence itself. This is the hallmark of the *a priori* method as explained by Engels. The concept, in this case the IQ test, becomes the object of study and speculation, while the reality, real intelligence, becomes more and more obscured by the theory claiming to grasp it. This process is partly explained by the early "success" of IQ tests. Results seem to confirm the *a priori* idea that intelligence is innate and fixed by birth. They seemed to justify the process of "sorting out" the "bright" from the "dull" at an early age and in general harmony with the already existing social hierarchy. It was the marriage of these *a priori,* ideological concepts and the established social relations that eventually gave birth to the IQ test.

IQ tests became fetishized and further theoretical development consisted in generalizations from the features of the test. Of course, the IQ test was not totally arbitrary, and reflected something in it. Any concept or abstraction from reality that arises in a practical situation usually contains some aspect of the reality it reflects. The essence of the dialectical materialist theory of reflection consists in recognizing that concepts and methods of study must constantly be measured against the reality that is being studied.[14] If a situation arises in which the reality is subordinated to the concept of it—

however appealing and pragmatically useful the concept or method—scientific development is inevitably replaced by metaphysical speculation. Rubbing sticks together was a very "successful" way of making fire. But if we defined fire by this particular way of making it, we would have a theory of combustion that would include, among other things, the pointedness of the sticks, the dryness of the leaves, the necessary age ranges of the fire makers, etc. Following the *a priori* method, we might study these properties intently, continually refining our "theory" by the use of sophisticated statistical techniques for showing correlations among all these properties, in order to derive a general concept of fire.

Jensen's assertion that we don't need to know what intelligence itself is does not really conflict with his assertion that IQ tests measure intelligence as an "objective reality" to the same extent that "atoms" or "genes" are an "aspect of objective reality." For "atoms" and "genes" are concepts in the same way in which "heat" is a concept. And it is possible to assert in the same breath that they are an "aspect of objective reality" as is any concept. In this sense "hobgoblins" too constitute an "aspect of objective reality" inasmuch as there are people who really believe in hobgoblins. But there is an essential difference between an imaginary concept and a real one—i.e., one that reflects something that really exists.

Relying upon an *a priori* subjectivist method, Jensen attempts to deduce "reality" from the properties of concepts, without having to raise any questions regarding something "out there," something that intelligence "really" is. The subjectivism inherent in this approach has led Jensen to defend it against the charge that IQ tests are completely arbitrary or "rigged," to suit the unfounded presuppositions of the test makers. This charge is made especially in relation to the obvious cultural bias of many IQ test items. After asserting that there is no answer to the question "as to what intelligence *really* is," Jensen tries to escape from the conclusion that it is therefore a figment of the psychologist's imagination with the following assertion: "The best we can do is to obtain measurements of certain kinds of behavior and look at their relationships to other phenomena and see if these relationships make any kind of sense and order. It is from these orderly relationships that we can gain some understanding of the phenomena."[15] Such an approach to "hobgoblins" might also find "orderly relationships" and correlations between phenomena, etc. But the orderly relationships of this kind do not demonstrate the real existence of hobgoblins.

Despite these initial theoretical criticisms, we will examine the "operations" whereby intelligence is defined. We will examine the "relationships" which Jensen finds between the phenomena. In particular we will look at the methods of comparison and "correlation" of data which give IQ a certain stability and so the semblance of objectivity. It is sufficient to say at this point that an agreement between data analyzed and measured by the same methods and within the framework of the same general concepts does not make the results more objective. Continual repetition of the same mistakes does not make them correct. Moreover, there is undoubtedly some "thing" or aspect of the phenomena of intelligence that is reflected by the IQ method, which may give underlying consistency to the results of analysis without there being an adequate explanation of what this is. The ideological or *a priori* method contains these two aspects, as Engels writes: "first, the meagre residue of real content which may possibly survive in those abstractions from which he starts and, secondly, the content which our ideologist once more introduces into it from his own consciousness."[16] Assuming that we find consistency in findings which result from the methodology of IQ analysis, we are not justified in concluding that we have explained this consistency and have thereby established real scientific understanding. Our task here will be to try to disentangle the degree of truth captured by the IQ method from the concepts and ideology that are imported into the reality from the consciousness of the mental testers themselves.

# 3

## Some IQ questions

Accepting for the moment the notion that "intelligence is what intelligence tests test," perhaps we will get some impression of what "intelligence" is and how it is measured by looking at some of the different kinds of items found in IQ tests. There we would expect to find some indication as to what IQ tests test, and as to whether this is in fact a fixed, innate capacity to think, to learn, to solve abstract problems.

### Examples of IQ test items

There are two main types of tests, individual tests and group tests. Individual tests are given to a single child by a specially trained examiner. The best known of these is the Stanford-Binet test. Group tests are devised so as to be given to a large group of children at one time, by an examiner who only has to read precise directions.

The earliest recognized "successful" intelligence test was developed by the French psychologist, Alfred Binet. The 1908 Simon-Binet test contained the following items, listed in the order of the ages at which the "average child" of that age was supposed to successfully complete them:

Age three:

1 - Points to nose, eyes, mouth.
2 - Repeats sentences of six syllables.
3 - Repeats two digits.

29

4 - Enumerates objects in a picture.
5 - Gives family name.

Age six:

1  Knows right and left; indicated by showing right hand and left ear.
2 - Repeats sentence of sixteen syllables.
3 - Chooses the prettier in each of three pairs of faces (esthetic comparison).
4 - Defines objects in terms of use.
5 - Executes three commissions.
6 - Knows own age.
7 - Knows morning and afternoon.

Age nine:

1 - Knows the date: day of week, day of month, month of year.
2 - Recites days of week.
3 - Makes change: four cents out of twenty in play-store transaction.
4 - Gives definitions which are superior to use; familiar objects are employed.
5 - Reads a passage and remembers six items.
6 - Arranges five equal-appearing cubes in order of weight.

Age twelve:

1 - Repeats seven digits.
2 - Gives three rhymes to a word (in one minute).
3 - Repeats a sentence of twenty-six syllables.
4 - Answers problem questions.
5 - Interprets pictures (as contrasted with simple descriptions).[1]

The 1960 Stanford-Binet Intelligence Scale is also presented in the form of items which are typical for a given age group. The test is more refined, with exactly six items and one alternate for each age level.[2]

At age three children are expected to:

1 - String four beads in two minutes.
2 - Identify ten pictures of the following objects by name: airplane, telephone, hat, ball, tree, key, horse, knife, coat, ship, umbrella,

foot, flag, cane, arm, pocket knife, pitcher, leaf; definitions by
description or by use, rather than by the correct or approximately
correct name, are scored as minuses.

3 - Build a bridge with blocks, in imitation of the tester; negative
score if the base blocks are touching or if the bridge falls.

4 - Identify from memory an animal which has been previously
identified in a picture; a minus score is given if the child goes on
to name other objects than the one which the tester had first
pointed out.

5 - Copy a circle; manual gives examples of scrawlings which are
correct and incorrect, leaving a certain amount of discretion to
the judgment of the scorer.

6 - Draw a vertical line; similar to the above, but only one example is
given and one trial, while three are allowed for the circle.

Alternate - Repeat three digits, pronounced distinctly "and with
perfectly uniform emphasis at the rate of one per second"; three
series are given.

For age six, the following items are considered to "measure
intelligence":

1 - Vocabulary. Six words should be defined correctly from the
vocabulary list of forty-five words. Eight-year-olds are expected
to define eight, ten-year-olds eleven, twelve-year-olds fifteen, and
fourteen-year-olds seventeen. The word list consists of the following forty-five words: orange, envelope, straw, puddle, tap, gown,
roar, eyelash, mars, juggler, scorch, lecture, skill, brunette,
muzzle, haste, peculiarity, priceless, regard, tolerate, dispropor-
tionate, lotus, shrewd, mosaic, stave, bewail, ochre, repose, am-
bergris, limpet, frustrate, flaunt, incrustation, retroactive,
philanthropy, piscatorial, milksop, harpy, depredation, perfunc-
tory, achromatic, casuistry, homonculus, sudorific, parterre. For
the scorer, examples are given of good and bad definitions.

2 - Differences. The child must know the difference between a bird
and a dog, a slipper and a boot, and wood and glass. A maximum
of two points is needed—so that one incorrect explanation would
not count against the child.

3 - Mutilated pictures. The child must verbally indicate missing
parts in pictures of a wagon, shoe, teapot, rabbit, and glove.

4 - Number concepts. The child must place before the examiner the

correct number of blocks requested, four out of five times. The examiner is carefully instructed not to help the child inadvertantly.

5 - Opposite analogies. The child must complete three of the four correctly: A table is made of wood; a window of . . . ; A bird flies; a fish . . . ; The point of a cane is blunt; the point of a knife is . . . ; An inch is short; a mile is. . . . For the last item the correct answers are listed as "long" and "longer," while "too long" and "long ways" are incorrect.

6 - Maze tracing. The child has to find the shortest paths in two mazes.

Alternate - Response to pictures. A picture is presented to the child, who is asked to "look at this picture and tell me all about it."

At age nine children are asked:

1 - To make a drawing of the figures formed by cutting a square in the side of a folded sheet of paper and a corner from the inside edge of a folded sheet of paper—without seeing the unfolded sheet.

2 - To explain "what is foolish" about three out of five "verbal absurdities." Eg.: "Bill Jones's feet are so big that he has to pull his trousers on over his head"; "One day we saw several icebergs that had been entirely melted by the warmth of the Gulf Stream."

3 - To draw from memory two designs from a card which has been exposed for ten seconds.

4 - To name a color that rhymes with head, a number that rhymes with tree, an animal that rhymes with fair, and a flower that rhymes with nose.

5 - To give the correct answer to two of three arithmetic problems given in the form of "making change." The child must give the right answer by calculating "in his head." ("If I were to buy four cents worth of candy and should give the storekeeper ten cents, how much money would I get back?" Same for twelve-fifteen and four-twenty-five.)

6 - To repeat four digits in reverse order to that stated by the examiner—three times.

Alternate - To give three words that rhyme with "head" and with "cap," respectively, given thirty seconds for each set.

At age twelve, the child must pass items involving:

1 - Vocabulary—fifteen of the word list referred to for age six.
2 - Verbal absurdities—answering four of the same five given at age nine.
3 - Picture absurdities, in which the examiner asks "What is foolish about that picture?" If the answer is "ambiguous," the examiner should rephrase the question to "Why is that foolish?" (Use of the word "foolish" seems to be obligatory for absurdities for eight-year-olds and older—while the examiner should say "funny" in the test for seven-year-olds.)
4 - Repeating five digits in reverse order—three series, only one for each.
5 - Definition of "abstract words"—pity, curiosity, grief, surprise. The procedure is to say "What do we mean by . . . ?" or "What is . . . ?" If the answer is ambiguous, further "clarification" is to be gotten by asking, "Yes, but what do we mean by . . . ?" or "What is . . . ?" The manual lists a variety of correct and incorrect or inadequate answers. Thus for "curiosity," "plus" is given for "Wondering what it is," and "Means that you are nosey," while "minus" is given for "Not to know" and "Somebody is always butting into someone else's business."
6 - To insert the appropriate word in the blank space for four sentences. Eg., "We like to pop corn . . . to roast chestnuts over the fire." The correct words are "and" and "or."
Alternate - To draw a design, more complicated than the earlier ones, from memory, aften ten seconds of exposure.

## Questions about these IQ test items

The above examples show the broad range of kinds of items that are used in standard intelligence tests. One justification given for such a variety of items is that "intelligence" is assumed to be a highly complex function, and so is measured by testing various abilities, information, and what seems to be just plain opinion (as "aesthetic preference"). This variety of types of items has given rise to frequent criticisms of IQ tests for being biased in favor of a particular culture or class. One critic, Jerome Kagan, analyzes the following "everyday problem situation" which a seven year old child is expected to solve: "What should you do if you were sent to buy a loaf of bread and the grocer said he didn't have any more?" Maximum credit is given to the reply, "I would go to another store." No credit is given for the answer, "Go home." The "everyday situation" for which the first is an "intelligent" response is, however, a very definite kind of situation. It is one in which there is more than one store within safe walking distance from home, where the child's family is not dependent on the particular store for credit, or where for any number of reasons the parent has not already explicitly instructed the child to go home in such a case. Kagan concludes:

> It is not surprising that rural and ghetto children are less likely to offer that answer. Recently I examined a set of protocols gathered on poor black children living in a large Eastern city and found that many of them answered the question by saying they would "go home"—a perfectly reasonable, even intelligent, answer for which they were not given credit.[3]

Difficulties of a similar type occur as well with respect to vocabulary tests, and with analogies which make use of words and concepts which may not be part of the normal experience of children in all classes and national or racial groups in our society. As Kagan notes, "The child is asked how a piano and a violin are alike, not how tortilla and frijole are similar."[4]

No doubt arithmetic tests test a child's actual ability to do arithmetic. But this is certainly a learned ability which depends upon complex and varying circumstances. How can we conclude from a child's actual performance on such tests that the item measures an innate, general mental capacity?

Reading comprehension tests measure vocabulary, reading ability, concentration, memory and other factors including "comprehension" of a situation which is described in a literary form. Here perhaps there is more ambiguity as to whether such items truly test some innate "intelligence." We have a tendency to think that ordinary practical "comprehension" is not "learned" but is rather something that occurs spontaneously. The fact that "reasoning ability" seems to be something unlearned, and presupposed to learning, gives such items the semblance of tapping an unlearned and so innate capacity.

Of course the same objection which Kagan gave to the problem which the seven-year old has to solve can be given here in a somewhat different form. Not only does familiarity with vocabulary and types of situations described vary among children of different social classes or with different national or racial background, but so does familiarity with reading and with different types of literature. Does the fact that a child's family environment does not include a regular emphasis on reading literature imply a limited *innate* capacity to "comprehend" literature?

This question raises a more complex problem, which is very important for the understanding of IQ tests. Is "comprehension" really an unlearned ability, a sort of "knack" which some children have to a greater degree than others and which explains their different levels of performance? Or is this too not learned, or developed, in the practical and scholastic activities of the child— although in a less explicit fashion than is the case with mathematics?

While critics of IQ tests can find examples of biased items, and raise deeper questions regarding the nature of general learning, the defenders of the thesis that IQ tests measure an innate capacity to think abstractly focus on items that seem to be the least affected by variations in the social environment of the child. Eysenck gives little indication of the more obviously problematic items included in many intelligence tests. Instead, he offers the following example to distinguish between an "achievement" test (A) and a test of mental "aptitude" or innate intelligence (B):[5]

A. Jupiter is to Mars as Zeus is to: Poseidon, Ares, Apollo, Hermes

B. Black is to white as high is to: green, tall, low, grey

Surely, Eysenck basically argues, no one would claim that the meanings of "black" and "white" are known only to certain classes or racial groups, as may be the case for the first example. The ability to solve this analogy is therefore clearly a matter of the exercise of an ability which requires no specialized learning.

Does this fact, if it is a fact, lead to the conclusion that the mental operation involved here is an expression of the child's innate capacity to think? In his own book of "do it yourself" IQ tests, Eysenck has concentrated on the more seemingly "culture-fair," unbiased sort of test.[6] The tests include "matrices"—where it is necessary to recognize formal patterns to complete a series of nonverbal figures. Eysenck seems to think that while one must "learn" who Poseidon was, and do so in a special form, one does not learn either to think in terms of formal opposites or to decipher varying degrees of complex formal patterns. But are these mental activities not also learned, and with varying degrees of facility, depending on complex circumstances? And if they are not explicitly taught in the schools, may they not be taught indirectly, say through the teaching of arithmetic? Are they incapable of being taught? In fact more advanced curriculums today are beginning to include exercise in precisely the types of formal thinking that may have previously seemed to have been developed spontaneously from the innate resources of the child.

In presenting his example of a good IQ test item, Eysenck neglects to mention that such an item is only "good" for a certain age group. While no four year olds may be able to get the right answer, most seven year olds may have no difficulty. The item might therefore be a "good" item for children who were six years old. To be "intelligent" means to pass certain tests at the right age. The "slow learner" is thereby identified as unintelligent or "incapable," even though he may eventually learn everything learned by the fast, "intelligent," learner. This author has found that the "verbal absurdities" listed for nine and twelve year olds generally evoke laughter when read to college students. If you see the absurdity, I ask, should not all of you get credit for being intelligent? Not, I reply, unless you laughed when you were nine years old or younger. Now it is already "too late."

This example illustrates an important peculiarity and limitation to the concept of intelligence implied by the IQ test. This peculiarity of the "definition" of intelligence implied in IQ tests is further seen in the fact that an average twelve year old cannot be considered to be more intelligent than an average six year old. The concept of intelligence is defined by the limited and *relative* concept of "being intelligent for one's age." While the twelve year old obviously knows more and has more developed mental abilities than a six year old, the first may be just as average for his own age as the second is for the group of six year olds. Because of this fact the IQ for each child is the same, and, for Jensen and Eysenck at least, they can be said to have about the same innate mental capacity.

In reply to this approach to intelligence, one can ask whether such a limitation of the definition of intelligence is not almost arbitrary. Why should not the older child, who can pass the more

'ficult tests, be considered more intelligent? Why should the child
) remains average at each age level, but is continually increasing
knowledge and mental abilities, be regarded as having the same
elligence and a certain fixed mental capacity?

y a brief examination of the types of items included in IQ tests
ave found that it is at least questionable whether such items
id on some innate capacity, unaffected by ordinary learning or
lying it. If the unlearned or innate character of the abilities
n such tests is more than questionable, so too is the supposedly
d character of the mental capacity which IQ tests are said to
easure. We note that the items are progressively more difficult for
older age groups. If such items measure intelligence, is this not an
indication that intelligence grows as the child grows? Is this not a
confirmation of the developmental character of intelligence? How is
it then that a child's intelligence can be defined as basically fixed?

The answer to these questions given by psychometricians, and
especially by Jensen and other contemporary hereditarians, will
require a more detailed examination of the way in which IQ tests are
constructed and general IQ theory.

By examining the kinds of items found in IQ tests we may have
dispelled somewhat the "magical aura" surrounding the concept of
IQ. It is less mysterious to know that one's IQ is determined by such
activities as the solving of problems in arithmetic. But it has perhaps become even more mysterious why being good or bad at
arithmetic, at such and such an age, should be considered a sign of
innate, fixed mental capacity—to say nothing of why the "correct"
answer to such items as "Which is prettier?" should even hint at
such a momentous fact. The examination of the items used in IQ
tests has in fact only confirmed our common sense as well as
philosophical opposition to the general theory of intelligence defended by Jensen and others. The items seem essentially to measure
"learned intelligence" rather than an inner, innate capacity. Moreover, the items seem to reflect a developing intelligence, rather than
a fixed capacity.

The main question that confronts us is therefore how, from
items that indicate both learning and development, can one deduce
an unlearned, inner "capacity" which remains the same throughout
one's life? To answer this question we need to turn to psychometric
theory and to a study of the construction of IQ tests.

B
we h
dep
un
use
ixe

4

*History of IQ theory*

...derstand more about why such items are said to measure
...gence" it is necessary to examine not only the *content* of the
but the *method* used in selecting these items and interpreting
scores. It is by examining the way in which test items are
...ted and the method of constructing the test as a whole that we
learn most about what an IQ is. At the same time, the *a priori*
...uppositions of IQ theory will stand out sharply.

The history of IQ testing does not show us a scientific attempt to
...erstand the nature of intelligence, its laws or dynamics. Instead,
...re is a consistent attempt from the beginning of the intelligence
...ing movement to develop a test that would make certain *a priori*
...ncepts about intelligence seem plausible. From the beginning,
...ere was an attempt to prove that intelligence is a biologically fixed
...pacity which is found in different proportions throughout society,
...o that it can be said that the upper class is innately superior in
...telligence to the "lower" classes, and the white colonialists are
...nately superior in intelligence to the non-white races over which
...ey seek to preserve their rule.

### *father of eugenics*

...to R. J. Herrnstein, the measurement of intelligence
... work of Francis Galton, a cousin of Charles Darwin
... *Hereditary Genius* (1869). Galton attempted to apply
...heory of biological evolution to the evolution of human

*39*

society and its stratification into higher and lower classes, as well as into what he considered to be superior and inferior races. As a "Social Darwinist," Galton believed that the eminent families of Great Britain had been selected in the societal struggle for survival of the fittest because of biologically superior traits. As a consequence of this belief, Galton became the founder of eugenics, the attempt to improve the quality of "the race" by biological methods and to prevent "inferior races and classes" from bringing down the quality of the human race as a whole. Herrnstein writes that "The inheritance of human capacity implied 'the practicability of supplanting inefficient human stock by better strains' and led Galton 'to consider whether it might not be our duty to do so by such effort as may be reasonable, thus exerting ourselves to further the ends of evolution more rapidly and with less distress than if events were left to their own course.'"[1]

Galton's ideological biases regarding human nature and the causes of class society are reflected in the way in which he set about attempting to measure intelligence. Above all the *a priori* character of his approach to the measurement of intelligence is implied in the fact that his tests, by his own admission, failed to measure intelligence.

Galton devised a number of tests designed to measure intelligence by testing the sensory reactions of people. Herrnstein mentions Galton's "ingenious tests for getting at sensory acuteness—measuring the ability to distinguish between weights differing only slightly, to hear tones of high pitch, to detect heat and cold, to feel a pin and so on."[2] Herrnstein writes that Galton's "hunch" that such tests would measure intelligence was proved wrong (although Herrnstein does not explain clearly why this was the case). To understand Galton's failure is to understand some of the basic ideas guiding the construction of intelligence tests which would eventually find a "successful" embodiment with the tests of Binet.

Why, it is necessary to ask, should the ability "to distinguish between weights differing only slightly" have been considered a way of measuring intelligence? The first point that stands out, when we compare this type of test with those now figuring in intelligence tests, is that Galton thought that intelligence would be reflected in *physical* abilities—in "physical intelligence." The idea that intelligence is a biologically innate capacity would be more plausible to common sense if it could be shown that seemingly unlearned physical responses were directly related to differences in intelligence. The problems that we raised regarding the acquired and developmental character of the abilities measured in present IQ tests would be less obvious were the tests to consist in "unlearned" physical reflexes.

Why, then, did Galton, and the intelligence test movement, not stay on this track? The answer is that the tests which Galton devised failed to demonstrate that people he *already* regarded as more intelligent—namely the British ruling families—were any more "intelligent" than common laborers. Because Galton assumed that intelligence exists in greater proportions among the wealthier and more powerful members of society (as well as those who were engaged in intellectual professions and had the "proper" schooling), he could not simply assert that doing better on his tests indicated greater intelligence. The results of the tests had to fit his presuppositions and they failed to do so. Consequently, it was the test and not the presuppositions that had to be changed.

Despite the failure of the particular tests devised by Galton, certain basic features of his approach continue to be at the foundation of intelligence measurement today. In the first place, a successful test is one whose results confirm judgments *already made* regarding which individuals are more intelligent—judgments made independently of the tests. The manner in which these judgments are made is only somewhat more sophisticated in hiding its assumptions than more open assertions of Galton.

Secondly, the nature of the type of "measurement" itself remains essentially the same. Intelligence measurements are comparative or relative, not direct and "absolute." The aim of Galton's test was not to measure the degree to which an individual possessed a *definite* ability, such as, for example, the ability to distinguish between two ounces of weight. What mattered to Galton was not the measurement of a definite ability but whether one individual had more of this ability than another. In setting up his tests for sensory acuteness, Galton had no way of knowing what a good performance was in itself. He simply found a range of performances, some inevitably better than others, some high, some low, and most grouped around the average.

The kind of measurement employed was a comparative statistical measure in which one's performance was compared with the average performance for the group of individuals taking the test. It is essential to understand that the concept of intelligence "measurement" refers to the statistical measurement of performances *relative* to each other. There is a basic distinction between the ordinary measurement of quantities and the statistical measurement of relative standing in a group. It is one thing to say, for example, that a table is three feet high and another thing to say that a table is "high" or "higher than average." Understanding these two types of measurement is essential to avoiding basic, common confusions and misconceptions as to just exactly what IQ "measures."

A third feature of IQ method simply involves making explicit the connection between the first and second. A "good" test is one in which those who score better are also those who are considered to be "more intelligent." The test results, showing a rank order or a relative standing of the performers, must fit or match the rank order or relative standing of individuals judged to be brighter or duller than others. This "matching" of two orderings of groups is at the heart of statistical "correlation." Perfect correlation between two "variables" exists when the variation in one perfectly matches the variation in another. Thus scores on tests "vary" from high to low. Galton was looking for a clear correlation between scores on his tests and his assumptions of which individuals are more intelligent. Galton was convinced that "native intelligence" is moreover highly "correlated" with social-economic class and professional status.

Finally, it can be said that although the particular nature of the test items has been significantly changed, the general idea which led Galton to select sensory-motor discrimination tests has not been completely abandoned. As Eysenck's example of a good IQ item clearly shows, it is still necessary to render as plausible as possible the idea that such items test something "unlearned," something like "native intelligence."

### Alfred Binet

The next major step in the development of intelligence tests was taken by the French psychologist Alfred Binet. It is probably significant that Binet was not interested in measuring innate capacity and did not believe that his tests in fact did so. For Binet broke significantly from the approach developed by Galton by rejecting the narrow focus on sensory-motor functions. Binet argued that intelligence, however vaguely defined, should be regarded as involving those complex developed abilities which people normally associate with intelligence. Moreover, intelligence was something that evolved over time.

Herrnstein attributes Binet's success to his "pragmatism," which "directed him to tests that sorted people out—for, whatever intelligence is, it varies from person to person. The sensory data did not distinguish among people as sharply as common sense required for a test of intelligence."[3] As this remark implies, Binet remained within the framework established by Galton inasmuch as he continued to measure intelligence comparatively or relatively, by differentiating the "bright from the dull," by devising tests to sort people out. However, to sort people out the way "common sense required" meant to test intelligence on the basis of items which "common sense" asserted to be involved in intelligence. Moreover, this meant to use items which changed with time, for the types of items that discriminate "bright" from "dull" five year olds must be different from those that discriminate between or measure the relative standing of ten year olds. The principle of measuring intelligence of people in general by establishing their abilities relative to each other was further refined by Binet when he took as his norm of measurement the average performance *at each chronological age* of the child's development. Binet described his approach as follows:

> We have come to see that education is a question of adaptation, and that in order to adapt it to the needs of the child we must make ourselves thoroughly acquainted with his or her mental characteristics. The principle, therefore, that guided me when forming this new laboratory was the knowledge of the average state of development of children of all ages—an entirely new idea in pedagogics.
>
> What my assistants and I set out to find out, in a strictly scientific manner, was the physical and mental value of the average child at various ages. Once having discovered this, we drew up tables of averages. We are able, for instance, to say: "This boy's growth is retarded. Though twelve years of age, he has only the development of a child of nine. He will require special attention and special nourishment."[4]

It is clear from this passage that Binet was oriented toward a developmental conception of intelligence, as well as toward a scientific study of child development to provide a basis for effective pedagogical practices. The discrimination of "backward" children was understood in the framework of attempting to orient education toward overcoming the backwardness, rather than toward identifying biologically fixed differences.

Nevertheless, there are important limitations of this approach to intelligence which lead to the possibility of maintaining the biological interpretation of intelligence. Not the least of these is indicated by Herrnstein who points out that "of all the countless ways one may want to distinguish between smarter and duller people, it may not seem especially insightful to choose the simple fact that during the first fifteen years of life, age confers intelligence (on the average)."[5] The use of chronological age as a basic ingredient of the definition of intelligence preserves the possibility of giving a biological, maturational interpretation of the growth of intelligence. Furthermore, the relativistic definition of intelligence by comparison with the average performance of one's own age group cancels out, in a way, the notion of development implied in the selection of progressively more difficult items. To the extent that one's intelligence is *defined* by one's relation to the performance of one's own age group, it becomes impossible to say that an older child is normally more intelligent than a younger one. If one's relation to the average performance for one's age does not change, then even though one learns more and acquires new skills every year, one's "intelligence" so defined will not change.

Finally, it is necessary to point to a fundamental weakness due to the "common sense" and "pragmatic" or empiricist approach of Binet. However much in advance of Galton's more evidently dogmatic approach, Binet's theory remains essentially unscientific and subject to ideological interpretations quite alien to his own. Binet could only protest against the hereditarian interpretations of his own work: " ... some recent philosophers appear to have given their moral support to the deplorable verdict that the intelligence of an individual is a fixed quantity. ... We must protest and act against this brutal pessimism."[6]

Binet's pragmatic, common sense approach to the construction of intelligence tests led him to incorporate into the method of test construction a basic assumption of Galton. Galton's tests failed to confirm or correlate with prior judgments regarding who was more intelligent than whom. Binet made sure that such a failure would not occur for his tests by building this prior judgment into the construction of the tests themselves. Not only did he change the types of items to include those which "common sense" identifies with intelligence, but he also allowed "common sense" to determine *who* was intelligent and systematically excluded items which did not confirm this prior judgment.

Binet set out "in a strictly scientific manner," as he says, to find "the physical and mental value of the average child at various ages." To make sure that his tests reflected "the average child" Binet simply asked teachers and school principals to select children at each grade level who were, in their opinion, average ones. He then found items which such "average children" could answer, with the restriction that the average children one year younger would not as a rule be able to correctly answer such items. The result was a progressively more difficult test with scores identified with average performances for each age. An alternative method used by Binet was to have teachers select the "brightest" and the "dullest" children in a given classroom. Items which discriminate between these two groups—that is, which the bright children can answer, but the dull children fail—are regarded as good intelligence items.

However pragmatically useful this method may be it is clear that it systematically builds into the test a "common sense," subjective judgment regarding who are the more of less intelligent people. Instead of the personal bias of the test constructor we have the more "impersonal" biases of the teachers in the school system. Intelligence is therefore defined in relation to the complex of abilities and attitudes which distinguish the "bright" from the "dull" in school work, at each age level, and in the opinion of teachers.

This approach excludes any *direct* relation to social class background, and is not so directly a reflection of class preferences as was true in the case of Galton. However, it indirectly builds into the test a class bias to the extent that the scholastic abilities and attitudes of the "successful" school performer tend to be fostered by the social and educational conditions of the children of middle class and professional families to a greater extent than among children of working class families, to say nothing of children from certain immigrant and from racially and nationally oppressed peoples.

The intelligence test developed by Binet is therefore essentially a *reflection* of the success of children in school work. The items selected for this test are systematically made to fit the prior judgment of who is more intelligent than whom. This judgment is essentially that of teachers and is based on their assessment of what constitutes a successful performance in school. In this respect Binet's test was unlike that of Galton in that Galton established tests of "intelligence" independently of his subjective and ideological assessment of which individuals were more intelligent. Binet's test, on the contrary, was *based on* common sense judgments of intelligence and so "success" was built into them.

The fact that intelligence tests generally reflect the rank order of children in school at each age group is the *basic feature* of these tests. The meaning and utility of these tests must be examined in the light of this feature. Interpretations of "intelligence scores" as reflections of innate capacity do not follow from the method of construction of the Binet test. Other features of intelligence scores will be examined shortly, with this "kernel" in mind. Thus the fact that "intelligence scores" generally remain stable is no proof of the existence of a fixed capacity. It is a registration of the fact that a child's "rank order" in school remains relatively stable. The *explanation* of this fact cannot be simply deduced from the stability of "intelligence," since "intelligence" is basically a description of the child's relative standing in school.

## Lewis Terman and the Stanford-Binet Intelligence Scale

As a result of his "successfully" constructed test, Binet could say, on the basis of a reasonably short test, that a certain child who was in fact six years old had the "intelligence" of an eight year old. Binet regarded this child as having an intelligence of "+2" or as being two years above average for his age. This relation between what was to be called the "mental age" of the child and the chronological age became the basis of what Lewis Terman at Stanford University in the United States would call the "Intelligence Quotient."

Terman adapted the Binet test to American conditions and refined the procedures for constructing tests, for selecting items and for making the tests representative of the population. Terman, along with fellow mental testers at the time, had no doubts as to the significance of the results of their tests. The contrast with the viewpoint of Binet is striking. Writing of the significance of "IQ" scores in the 70–80 range, Terman wrote (1917) that this level of intelligence

> is very, very common among Spanish-Indian and Mexican famil-
> ies of the Southwest and also among negroes. Their dullness
> seems to be racial, or at least inherent in the family stocks from
> which they come . . . the whole question of racial differences in
> mental traits will have to be taken up anew and by experimental
> methods. The writer predicts that when this is done there will be
> discovered enormously significant racial differences in general
> intelligence, differences which cannot be wiped out by any
> scheme of mental culture.
>
> Children of this group should be segregated in special classes
> . . . They cannot master abstractions, but they can often be made
> efficient workers . . . There is no possibility at present of convinc-
> ing society that they should not be allowed to reproduce, although
> from a eugenic point of view they constitute a grave problem
> because of their unusually prolific breeding.[7]

The founder of the most important mental test, the Stanford-
Binet Intelligence Scale (1917), thus shared the main ideas of Galton.
Gone is the purely pragmatic approach of Binet, who saw his tests as
practical instruments for identifying children who needed special
attention in order to overcome their educational deficiencies. Instead there is the identification of permanently defective individuals
and a veritable racial caste of manual laborers. In 1912, before
Terman's own modification of the Binet test, Henry Goddard, working for the United States Public Health Service, was using Binet's
test to identify "mental defectives" among immigrants landing at
Ellis Island in the New York harbor. Goddard did not hesitate to
proclaim that 83% of the Jews, 80% of the Hungarians, 79% of the
Italians and 87% of the Russians were "feeble-minded."[8]

The hereditarian interpretation of Binet's tests thus stamped
them from the beginning of their life on the North American conti-
nent. Perhaps, one might suppose, there has been some additional
feature which Terman introduced into Binet's test that might have
partially excused his drawing such far-reaching conclusions as to
the mental capabilities of the most exploited and abused sectors of
the U.S. population.

In the first place, Terman devised the concept of the "Intelligence Quotient" or IQ. Instead of subtracting the chronological age of the child from the "mental age," Terman divided the latter by the former. Thus a child with a "mental age" score of 8 and a chronological age of 6 would score $8/6 = 1.33 \times 100$ (to eliminate the decimal point) or 133.

Secondly, Terman built into his test a more precise statistical requirement. The rank order of the children at any particular age should approximate the "normal curve." The normal curve has the following appearance:[9]

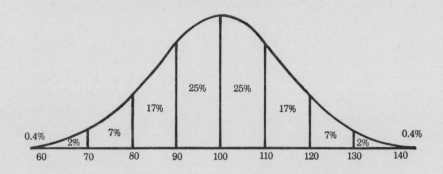

Whereas Binet used the rough estimate of teachers, in a relatively small sampling of the school population, as to what an average performance consisted in, Terman corrected the impreciseness of this method by using more exact statistical definitions and by improving the method of selecting and scoring items as well as of administering the tests. Terman criticized Binet's essentially two-valued standard. While Binet was primarily interested in vaguely distinguishing children who were "normal" from those who were significantly backward, Terman argued that "intelligence" involves a more finely graded range of capacities, as described by the normal curve.[10] Complex statistical methods for determining the "reliability" of tests as well as various methods of determining their "validity" add to the methodological and theoretical framework. It is these features probably more than anything else that lend to the IQ test its appearance of being a rigorous scientific instrument. As we will see, much of the current arguments that IQ measures innate capacity stem from speculation regarding the properties of these statistical methods used in the construction of the tests.

While the statistical requirement of the normal curve involves a refinement of Binet's principle, it does not substantially change it. While Binet was satisfied with a rough estimate of who were the average children in school work at each given age, Terman sought a more precise method of discriminating between children at any one chronological age. By requiring that the outcome of the scores fit the normal curve, the Stanford-Binet test gives the IQ an exact statistical meaning, which can be easily compared with other statistical measurements. Terman, however, not only constructed his test on the basis of certain statistical requirements, such as the normal curve distribution of the scores, he also believed that intelligence is inherently distributed "normally," i.e., with most scores clustering around the average, with progressively fewer individuals achieving higher or lower scores. As we will see, the interpretation of IQ as a measurement of fixed, innate intellectual capacity stems primarily from reading "absolute" and objective meaning into the methods adopted by test constructors—methods which may have a limited value, but which do not justify the extrapolations of the IQ theoreticians.

Despite the rigorous statistical form given to IQ tests, the construction of the test is a complex and largely empirical, trial and error process. It is necessary to construct a test whose results will 1) distribute scores for each age on a normal curve, 2) "correlate" or match with school performance as the main criterion of the success of the test, 3) establish this correlation for the population of the entire country, and 4) use test items which, although increasingly more difficult, nevertheless appear to be untaught or to involve "innate, general intelligence."

These objectives were accomplished by Terman in his 1916 test and its later revisions by a laborious process, which, as in the case of Binet, involved considerable trial and error.

In the first place, Terman constructed his test on a large, representative sample of white, native-born Americans. This sample involved 1,000 children of all school ages plus 400 adults, male and female, rural and urban, and from families of different incomes, in proportions that roughly approximated the national proportions. The test is originally constructed on a "standardization sample" of the population. If the standardization sample is selected carefully, the test should be representative of the "whole population." Thus, if the test finally produces a normal curve distribution for the standardization sample, the same results should occur when the test is then released for use in the entire population. Were the standardization sample overly representative of, say, urban, middle class children, a test which was constructed for them would result in a nonnormal distribution for the whole population which would show the majority of the population as "below average in intelligence." This is of course not a sign of defectiveness of the national intelligence, but of the defectiveness of the test. Because of the difficulties of selecting a really representative sample of the population, tests are always subject to a certain "sampling error." Were it to turn out, for example, that the majority of those taking the test were below average, adjustments could be made either in the placement of items in the test or in the scoring technique to bring the scores up to the normal distribution.[11]

It is important to note that the standardization sample selected by Terman consisted of "white, native-born Americans," and excluded non-white Americans and the foreign-born. This omission is especially striking in view of the interest demonstrated in examining the "native intelligence" of just such sections of the population. A test based on the white, native-born population would necessarily be unrepresentative of populations not included in the standardization sample.[12] Terman's reasoning in selecting this sample seems to be that he wished to create a test of "true American" intelligence, which would be the standard against which other peoples should be measured.

When the sample has been selected it is then necessary to select items which show the necessary "age discrimination." This requirement in itself creates considerable difficulty. There is no exact formula for determining what percentage of children of a certain age should pass an item for that item to be regarded as representative for that age.

Age discrimination is an important feature of item selection, but not the main criterion of the "validity" of the test. It is conceivable that a progressively more difficult test could be constructed which discriminates on the basis of age but which does not measure "intelligence." That is, such a test would not discriminate the "bright" from the "dull" and correlate with school performance and the opinions of "common sense." It is conceivable that a test using vocabulary and situations from popular sports might show the superiority of children who do poorly in school. Such items are eliminated from IQ tests by definition. The items used in the test should therefore be ones on which the "more intelligent" children do better. Again, it is not a question of devising a test on the basis of a theory of what constitutes intelligence, and then testing the children to see whether or not they are "intelligent" as independently defined on the basis of psychological theory. Galton's test failed with this approach to produce results which confirmed his assumptions of what the results should be. Binet's "pragmatic" approach was therefore adopted by Terman who selected his items so that children judged to be "bright" would do better than those judged to be "dull."[13] Thus Terman's Stanford-Binet has built into it a criterion of school success which can be said to be the main thing which the test measures. Any further notions that the test measures general intelligence or intellectual capacity is based on an extrapolation from this fact, as well as from the *a priori* statistical features of test construction.

One further point of major significance should be noted in regard to item selection. As tests of "general intelligence," IQ test items should not appear to involve specialized learning. The vocabulary and the "experiences" used in the test should be those which are in the range of everyone—at least of a given age. As a test that is meant to discriminate children on the basis of general intelligence or intellectual capacity, the tests should not appear to involve symbols with which the child is unfamiliar or mental operations which require special training. Thus, for example, questions involving analytic geometry would not be "valid" for use as measures of "general intelligence," because it is well known that not everyone past a certain age has had the "opportunity" or taken the opportunity of learning analytic geometry. This is not the case, on the other hand, for arithmetic, or, after a certain age, simple algebra. The assumption is that the symbols and operations of arithmetic are known to all children above a certain age, and that these "environmentally acquired" aspects of thought process are common. The success at such items will therefore differentiate between children, not on the basis of their environmental experience, but according to their "native intelligence."

This general theoretical and philosophical definition of intelligence as the inner, native capacity of thought, which assimilates the environmental "input," is clearly present in the example of a "good" item selected by Eysenck. However, the fact that some do better than others on a given item and that the better scorers are also those who are better in school is not the result of tests which are constructed on the basis of some theory of fairness, but simply a condition for the building of the intelligence test. Nevertheless, in the selection of items to meet these conditions, items are chosen which *seem* like items reflecting "general intellectual capacity" as opposed to special acquired "achievements." Consequently, IQ test items have a somewhat peculiar character. Analogies, opposites, comprehension, vocabulary, common sense opinions, etc., are all types of items which "seem to" meet this last condition. They are not *obviously* "achievement" tests. *Intelligence tests are therefore tests which reflect school performance by the use of items that do not obviously reflect school learning.* This too is a basic precondition of intelligence test construction. Nevertheless, it lends plausibility to the concept that what is being measured is not achievement or learning, but general intellectual capacity.

Other features of IQ tests will be discussed to the extent that they are important to the arguments of Jensen and others who see IQ as a measure of innate intellectual capacity. Our main objective is to examine this claim and to show that it is unjustified on both theoretical and empirical grounds. To do this we have already underlined some of the main presuppositions of IQ test. In the following pages these presuppositions will be critically examined. Psychometric theory involves an elaborate system of concepts and methods designed to justify the IQ test as a measure of "intelligence" which is both "reliable" and "valid." Close examination of this system shows, however, that it does not validate the notion of IQ as a measure of general intellectual capacity. At most it can be said that IQ tests measure rank order of school performance. Jensen, Eysenck, and Herrnstein argue nevertheless that IQ is a more profound measure of deeper intellectual capabilities. We will examine the arguments which attempt to support the classical hereditarian theory of intelligence measurement and show that in every case such arguments are based on *a priori* assumptions designed to make plausible the theory that at the basis of differences in intellectual performance are differences in innate capacity.

# 5

## "Reliability" and "validity" of IQ tests

Test manuals describe the degree of "reliability" and the "validity" of the tests, which may suggest that IQ tests are "reliable" measures of intellectual capacity and have been validated by some criteria independent of the *a priori* conditions of testing.

The less important for our purposes is the question of "reliability," although this may be quite important from a technical point of view. More important is the question of "validity," a concept that seems to be more related to the notion that the tests have somehow been *"proven"* to measure "intelligence."

In addition to the "reliability" of tests due to minimizing "sampling error," described above, there are different ways of estimating the ranges of error which may be attributed to a test as a result either of short-term fluctuations in the psychological state of the individual taking the test or of chance familiarity or ignorance of material involved in specific test items. The second main aspect of reliability has to do with the manner in which the test is administered.

### Reliability

Without going into detail regarding methods of determining reliability, it is enough to note that this concept refers to the elimination of short-term variations in the performance of the child or the

**53**

tester. In introducing the subject of "mental tests," one psychologist writes, "Mental testing is simply a more refined and scientific method of doing something which we all do every day of our lives, that is assessing one another's abilities and character traits... For example, we might watch a small boy playing with bricks, and from his skillfulness, or the elaborateness of the resulting construction, jump to the conclusion that he is quite a bright child. Obviously the basis for this and other such judgments is very haphazard and unreliable."[1]

It is one thing for an individual to guess the relative intelligence of a child, it is another matter for many individuals to make such a guess. IQ tests basically reflect the average opinion or "representative" judgment of teachers regarding the relative performance of a child compared to children of the same age for the entire nation. It is in this sense that the tests are more *reliable* than the snap judgment of a single person. But we would not agree that the tests are thereby more scientific. There may be continuity between the intuitive judgment of the individual and the collective judgment of society, but it is still a matter of a subjective judgment operating within the narrow constraints of a definition of intelligence which is itself far from complete and scientific.

Statistical methodology may make an opinion poll a far more "reliable" indicator of the general opinion of a nation than a brief conversation with one's next door neighbor. This does not make the opinions expressed through the opinion poll *truer* in themselves, although they may be more representative than the opinions of one's neighbor. This point is considerably obscured, however, in the case of the opinion polling regarding intelligence which is at the basis of IQ. The methodology used may make such results more representative and "reliable" than the snap judgment of an individual as a way of "measuring" the intelligence of a particular child. But when some psychologists boast that IQ tests are more reliable predictors of an individual's performance than are judgments based on particular interviews, it should not be thought that this fact, if it is generally true, proves that a scientific way of measuring general intellectual capacity has been discovered. IQ tests are only a "refined" form of the relative estimate of intelligence made by the common sense judgment of teachers and psychologists.[2]

## Validity

The term "validity" in psychometric theory implies more than the simple reliability of test scores. Not only are such scores freed from chance variations in the mood of the child or the tester, but they are said to be validated by some criterion *independent* of the test construction as actually measuring "intelligence." The independence of the criterion for validating a measurement is crucial to proving that one has not simply forced an object into the arbitrary confines of an *a priori* conception of intelligence and system of measuring it. It is often possible to read what we want to into a particular event or character. Often a person's appraisal of another person tells us more of the character of the first person than of the second. If the person who makes the judgment is consistent we can attribute a great deal of "reliability" to such judgments. We can rely on this person to make the same biased judgments in every case. But we recognize that this reliability does not make the judgments valid, i.e., *true.* The judgments are validated by some criterion that is *independent* of the mere fact of the judgment or of its short-term consistency. Can IQ tests be validated by some criterion that is independent of the fact that a concept and method of measuring IQ is applied consistently and "standardized?"

Four ways of determining validity are generally recognized: content, predictive, concurrent and construct validity.[3] A brief description and commentary on these four ways of validating tests will further clarify the nature of IQ tests and the technical theory of test construction. We will see that none of these criteria can truly be said to be independent of the way tests are constructed and to prove that the tests in fact measure "general cognitive ability."

### Content validity

"Content validity" refers to the problem of whether a test "covers a representative sample of the behavior domain to be measured."[4] Thus a test that is supposed to measure musical ability should be composed of more than tests of an individual's knowledge of major composers and their works. To measure an individual's musical ability it is necessary to understand various relevant aspects of this ability, and provide tests which demonstrate competency in each of these aspects.

The concept of content validity implies that one understands the ability in question and can analyze its component parts in such a way as to construct a test that accurately "samples" the various features of that ability. Real demonstration of the ability, in the degree predicted, would seem to be the ultimate criterion of the validity of the test. In the case of the IQ test, however, there is no generally agreed definition of the area of "general cognitive ability," if that is what the IQ is supposed to measure. Thus the content of the tests was not derived by an analysis of the "behavior-domain," general cognitive ability, but simply by selecting items that discriminate children on the basis of their school performance but which do not appear to be directly learned through school. The "content validity" seems to consist in the fact that the items look like items that measure "general intelligence" as defined by intuition.

But this is more a matter of giving plausibility to the test than of analyzing an objective domain into its component parts. Anne Anastasi notes that IQ tests have no clear validity on this score: "Aptitude and personality tests bear less intrinsic resemblance to the behavior domain they are trying to sample than do achievement tests. Consequently the content of aptitude and personality tests can do little more than reveal the hypotheses that led the test constructor to choose a certain type of content for measuring a specified trait. Such hypotheses need to be empirically confirmed to establish the validity of the test."[5] In other words, we need to look elsewhere to confirm the hypothesis which led to the selection of the types of items that appear in IQ tests.

Tests which effectively measure the existence of certain abilities, and are proven to be "valid" by actual performance involving the ability are still not scientific in the full sense of the term. David McClelland, who argues that IQ tests are simply measures of school standing and not of innate capacity, calls for tests based on real "competence." He hopes thereby to eliminate the invidiousness of so-called intelligence tests which assume general intellectual incapacity.[6] Testing for "competence" rather than "intelligence" may eliminate many of the abuses connected with IQ tests. However, as long as the causes of the development of the abilities in question are not explained, tests remain a matter of measuring already formed abilities. Structural biases in the development of "competencies" would still be reflected in the use of such tests. Moreover, hereditarian explanations of the more exactly defined "competence" are not eliminated by such reforms. McClelland attacks the hereditarian theory of IQ in a manner similar to Anastasi, i.e., *within* the framework of "sorting and selecting," but not of *explaining* development so as to be able to master the processes involved in various social and intellectual skills.

## Predictive validity

The second type of validity ascribed to IQ tests is *"predictive validity."* Of the various criteria of validity this one seems to be the most significant. As a general rule IQ scores are relatively stable. Especially after the age of six or so, an individual's IQ score remains relatively constant.[7] Consequently, IQ scores are good predictors of relative school performance. However it is not surprising that IQ tests predict school performance to a significant degree. This is only a reflection of the fact that once a rank order is established in any school it tends to be relatively stable. Children who are "at the top" in their early years tend to stay there, while children at the bottom remain there as well. This is true not only within schools, but, perhaps even more importantly, between schools. Good schools generally remain good schools, and poor schools stay poor. If IQ tests are constructed to reflect the rank order of performances at each age level in relation to a national average, there is no mystery to the relative stability of IQs and to the predictive power of IQ tests regarding school achievement. This is nothing other than a reflection of a well-known fact of life in the schools—as well as between the schools. This stability of the rank order of school performances is therefore not *explained* by IQ, since the stability of IQ is basically a reflection of this stability of relative performance.

It is not necessary to hypothesize differences in fixed mental capacity to explain the degree of fixity of rank order in school success. Differentiation between children is not simply a spontaneous occurrence in schools where everyone is "treated equally." Not only are there major differences between schools, but within schools themselves there is a *system* of "tracking" or "streaming" children which tends to guarantee a stability of rank order.

At the most general level, differentiation between individuals in school is guaranteed by a system of instruction which places primary emphasis on competition between individuals—rather than, for example, on cooperative efforts and joint projects of children, and/or competition between groups of children.[8]

Closely bound up with the general feature of individual competition in the classrooms is the tracking of children in levels of achievement. The tracking system tends to rigidify rank order. Although tracking is often justified on the ground that children have different basic capacities, this justification and explanation is more in the order of a self-fulfilling prophecy—as is the predictability of IQ tests. Even if IQ scores themselves play no role in determining expectations—both in the teacher and quite soon in the pupil as well—relative judgments of individual standing would nevertheless form and solidify the rank order of children in schools. Common practices, such as placing the best readers in the first row, combined with disciplinary measures, rewards and punishments, etc., have substantial influence over an individual's eventual performance and "self-image." Deeper studies of the psychological and pedagogical effects of individual competiton and rank ordering according to general levels of performance would shed much light on the question of explaining relative stability of rank ordering.

Of primary importance in an examination of the dynamics of learning is the presence of racism, creating an oppressive climate for minority children and acting to undermine educational motivation for children, many of whom see little encouragement or opportunity to develop useful academic skills. The existence of a basically segregated school system in the United States constitutes a national tracking system in which race, and not IQ, is clearly of primary importance.

Segregated schooling, class differences between schools, the tracking system, and competitive evaluation of individual performance constitute a complex, practical system of education. This system is often justified on the grounds that there are innate, individual, class and racial differences in capacity. Since, it is argued, the schools provide "equal opportunity," the differences in performance must be due to differences in either "basic ability" or "effort" in individuals themselves. In fact the schools offer no such equal opportunity, but are themselves based on a concept of differential education. The fact that primary schools are more homogeneous in form than secondary schools gives the impression that for some eight years children are given an equal chance to "show what they are made of," and only with secondary school are they finally tracked along vocational or academic lines. A deeper analysis of practices on the primary school level shows that this "fairness" is superficial. All levels of education involve unequal relations between teachers and students, and between the children themselves as they internalize their own relative degrees of academic inferiority and superiority.

Thus "predictive validity" essentially raises more questions than it answers. It does not refute our claim that "intelligence" and the IQ methods for measuring it rest on certain presuppositions which are not independently verified. Since the process of verification itself remains within the sphere of these presuppositions, there is no proof by correlations with later school performance that IQ tests measure general intellectual capacity.

## Concurrent validity

The third form of "validity" is *concurrent validity*. This concept has to do with the validation of the test, either by the actual performance which the test is meant to measure, or by correlation of the scores with the scores of other tests meant to measure the same thing.

As to correlation with other test scores this involves an obvious begging of the question. For the question remains as to the basis of the validity of these other tests. The widespread practice of "validating" other tests, particularly group tests, by the correlation of scores with the Stanford-Binet is only a testimony to the importance of the latter. It hardly shows the validity of the other tests as measures of "general intellectual capacity."

Under the heading of concurrent validity Anastasi lists using academic achievement and years of schooling, teachers' ratings or the ratings of psychologists and trained observers, as well as the method of "contrasted groups." Under "ratings" Anastasi states that "personal judgments" can be the "very core of the criterion measures. Under these circumstances the ratings themselves define the criterion."[9] The method of contrasted groups is simply the method of selecting "bright" and "dull" groups of children, and measuring individual items and the test as a whole as to their "validity" in distinguishing these opposed groups. Anastasi stresses the use of groups which have been sorted out by "the cumulative and uncontrolled selective influences of everyday life.... The criterion is ultimately based upon survival within a particular group versus elimination therefrom."[10] An example of this is the contrast of "institutionalized mental defectives" and normal school children of the same age. (Presumably, children in the first group have not been institutionalized as a result of their performance on IQ tests.)

The use of "survival of the fittest" theory to give validity to IQ tests harkens back to Social Darwinism. In developing tests of adult intelligence Terman used "contrasting groups" of 30 business men, 150 "'migrating' unemployed men," 150 adolescent delinquents and 50 high school students. The assumption in this selection seems to be that "life itself" has selected these individuals on the basis of their innate intelligence. The tests should thus reflect the superiority of the businessmen over the unemployed, and the high school students over the "adolescent delinquents." Terman does not go into any detail regarding his method of selecting items, noting only that "Adults whose intelligence is *known from other sources to be superior* are found to test well up toward the 'superior adult level,' and this holds whether the subjects in question are well educated or practically unschooled. The almost entirely unschooled business men, in fact, tested fully as well as high-school juniors and seniors."[11] The fact that the "unschooled" businessmen scored as well as the high school student is obviously understood to mean that these tests reflect "native" intelligence.

Although no information is given here on how items are selected, it seems clear that this "finding" is actually a premise for the construction of tests meant to show the native intelligence of economic survivors to be superior to life's dropouts. Items could presumably be found which the delinquent adolescents and the migrating unemployed could answer, and which the unschooled businessmen and (their?) schooled adolescents could not.

Thus "concurrent validity" for the Stanford-Binet consists in the agreement of the test with judgments of intelligence made by school teachers, with survival in school versus dropping out, and success in life versus unemployment. But these criteria are a basis of test construction, not a confirmation of the validity of the test constructed on some other basis. The justification of this procedure is the *a priori* notion that success in school and life is determined by innate intelligence and therefore tests which distinguish the successful (in school and in life) from the unsuccessful will have measured innate intelligence.

## Construct validity

The fact that these three methods of validating IQ as a measure of general intelligence fail to do so leads to a forth, murkier, and more clearly subjective theory of validation—so-called *construct validity*.

According to Anastasi, "The construct validity of a test is the extent to which the test may be said to measure a 'theoretical' construct or trait." Such a "construct" as "intelligence" involves a "broader, more enduring and more abstract kind of behavioral description than the previously discussed types of validity" and so requires "the gradual accumulation of information from a variety of sources. Any data throwing light on the nature of the trait under consideration and the conditions affecting its development and manifestations are grist for this validity mill. As illustrations of the specific techniques utilized may be mentioned age differentiation, correlations with other tests, factor analysis, internal consistency, and effect of experimental variables on test scores."[12]

Essentially, the above description of construct validity underlines a kind of scientific limbo in which the problem of validation places IQ tests. Anything that seems to give some support to the vague concept of intelligence is grist for this last, catch-all validity mill. Jensen's method of proving the innateness of IQ consists precisely in accumulating "various lines of evidence, not one of which is definitive alone, but which viewed all together, make it a not unreasonable hypothesis that genetic factors are strongly implicated in the average Negro-white intelligence difference."[13]

There is no need to elaborate on the matter of age differentiation which is clearly a condition of the construction of the test, and not an independently discovered correlation. Anastasi admits that various correlations with other tests are similar to techniques discussed under concurrent and content validity.

How does "factor analysis" demonstrate the "validity" of the tests? According to Anastasi,[14] factor analysis first of all simplifies the test by breaking it down into a small number of categories from the originally large number of particular items. Each "factor"— such as vocabulary, verbal comprehension, analogies, etc.—is then correlated with the overall test in order to determine the "factorial validity" of that particular factor. Thus, for example, if children, who have an average IQ happen also to have average scores on the vocabulary tests, but above average (or below average) scores on analogies, the vocabulary subtest is said to have a higher factorial validity than the analogies subtest. If the score on the subtest is the same as the score on the entire test, the subtest has a factorial validity of 1.00—perfect correlation. If the factorial validity of a subtest is very low—showing little correlation with the test as a whole—it is then doubtful whether the subtest belongs in the test at all. Thus factorial validity is a method of validating particular groupings of items by comparing them with the test as a whole. Instead of a validating factor outside of and independent of the test, the test itself is here regarded as the criterion of the validity of the factors that compose it.

### Jensen's arguments for validity: "g" and Headstart

Jensen regards Charles Spearman's correlations of the subtests to be an especially cogent argument that the tests really measure a basic "source of variation" of performance, some "general intelligence," or simply "g." Thus, according to Jensen,

> Spearman noted that if the tests called for the operation of "higher mental processes," as opposed to sheer sensory acuity, reflex behavior, or the execution of established habits, they showed positive intercorrelations, although the tests bore no superficial resemblance to one another ... For example, a vocabulary test shows correlations in the range of .50 to .60 with a test that consists of copying sets of designs with colored blocks; and a test of general information correlates about .50 with a test that involves winding through a printed maze with a pencil. . . . To account for the intercorrelations of "mental" tests, he hypothesized the existence of a single factor common to all tests involving complex mental processes . . . Spearman called the common factor 'general intelligence' or simply g.[15]

Thus, for Jensen, factor analysis is a method of validating IQ tests because it breaks the test down into a number of different factors, and then demonstrates that all of these factors "intercorrelate" to a significant degree. Does this high intercorrelation prove that IQ tests measure "general intelligence"? Here again, a little reflection on the method of test construction shows that there is no independent criterion of validation of the IQ score. Intercorrelation of the items is in fact a condition for the construction of the test. Items which the "bright" children do not pass are not considered to be good items for distinguishing "bright" from "dull" children at a particular age. Such items are either eliminated or placed somewhere else on the age scale where they "work." Thus all the items of the test must by definition "intercorrelate" as much as is possible given the empirical, trial and error manner of test construction.

Jensen does not explain this matter of the *a priori* character of item intercorrelations. However, he makes a "statistical qualification" for the benefit of those who might understand that "g" is an artifact of test construction: "We should not reify g as an entity, of course, since it is only a hypothetical construct intended to explain covariation among tests. It is a hypothetical source of variance (individual differences) in test scores."[16] In fact "g" cannot explain anything. The fact of intercorrelations is itself explained by the method of test construction. It is Jensen who hypothesizes that the "hypothetical construct" explains differences in performance. In fact it only reflects these differences. Jensen quickly removes any doubt cast on "g" by his qualification of it as not an "entity," as a "hypothetical construct," etc: "Despite numerous theoretical attacks on Spearman's basic notion of a general factor, g has stood like a rock of Gibraltar in psychometrics, defying any attempt to construct a test of complex problem solving which excludes it."[17]

The *a priori* artificial character of "g" is brought out by Anastasi, when she examines "internal consistency" as a method of validating tests. Thus, in one method of constructing a test two groups of children are identified, one judged to be "bright," the other "dull":

> Items that fail to show a significantly greater proportion of "passes" in the upper than in the lower criterion group are considered invalid, are either eliminated or revised. Only those items yielding significant item-test correlations would be retained.[18]

Thus correlations between scores on the test as a whole and either individual items or groups of items is a necessary condition of test construction. The fact that factor analysis "finds" this general consistency in the test, despite the relative diversity of items which make up the test, is not at all surprising and does not demonstrate that the tests measure "general intelligence." At best it is a reflection of the fact that children who do better than others in school do so in a number of relatively different areas of mental activity.

As to "effect of experimental variables on test scores," this method of construct validation refers to experiments designed to confirm the validity of the scores by retesting under different conditions.

Experimental testing of the construct of IQ as a fixed intellectual capacity would consist in "interventions" in the normal school system to see whether such interventions would change IQ scores. Thus, if significant variation of the school system does not change the rank ordering of the children, then the concept that IQ measures a basically fixed capacity would then appear to be at least partially confirmed. When Jensen refers to Operation Headstart as a "failure," he is arguing that just such experimental variation of the school system had in fact occurred, and had failed to "boost" the IQs of the children involved. However, as Jensen himself admits, IQ scores were in fact affected positively by Headstart. Jensen argues that this boosting was only temporary and disappeared after two years.[19] However, the fact that after two years of returning to the "normal" situation children's IQs or rank order returned to "normal" does not mean that Headstart failed. Only if the intervention was supposed to affect a "general intelligence" or "g" that remains at the basis of all learning could it be said that Headstart failed to affect this "hypothetical construct." If IQ tests simply involve the testing of skills and ideas somehow involved in school performance at a *particular* age, an intervention program which increased performance in *those* skills would not directly affect the *different* abilities which represent normal performance at a later age.

Headstart only failed to confirm Jensen's own hypothesis that intelligence is an underlying "source of variance" of school performance, which if affected by an "environmental" effort should remain durably affected. Our argument, on the contrary, is that this definition of intelligence is theoretically and empirically unjustifiable. The fact that a temporary intervention should have only temporary success is hardly proof that it is a failure. The limited character of both the effort and the results naturally suggests the need for a more expanded program. It is in the face of this perspective that Jensen argues that the experiment failed to boost "intelligence"—an admittedly "hypothetical construct"—while "only" improving more superficial "achievement."

In conclusion, none of the standard arguments for validating IQ ⌐ proves anything other than that these scores reflect rank order in school performance and a relative stability of this rank order. There is no validation of the "constructs" "general intelligence," "cognitive capacity" or, most importantly, "innate, general cognitive capacity." All these "constructs" remain simply that—constructions of the mind, *a priori* ideas that have guided the development of a test that seems to make these ideas *plausible*. In the light of the above discussion it is unclear what the concept of "validity" really means for IQ tests. At best it seems to mean that the tests have in fact been constructed according to the rules laid down for their construction. Aside from this there appears to be no "validation" of these rules, and the concepts or hypotheses standing behind them, by a practical result which is not essentially presupposed by the test in the first place.

## Behaviorist interpretation

Although Anastasi, for example, seems to support the methodology of validation as demonstrating that IQ tests are valid measures of "something," her own arguments rule against the notion that any of the major concepts of intelligence testing are valid. Thus, in a discussion of construct validation, Anastasi argues that no hard and fast line can be drawn between construct validation and the other forms of validation. In fact the latter might all be included under the broadly defined notion of construct validation.

Anastasi concludes with a warning against speculation on the meaning of various kinds of tests:

> It would also seem desirable to retain the concept of the criterion in construct validation, not as a specific practical measure to be predicted, but more generally to refer to independently gathered *external data*. The need to base all validation on data, rather than

on armchair speculation, would thus be re-emphasized, as would the need for data external to the test scores themselves. Internal analysis of the test, through item-test correlations, factorial analyses of test items, etc., is never an adequate substitute for external validation.[20]

So far we have failed to find any external validation of IQ scores as measures of anything other than rank order of school performance, which is what IQ tests are based on in the first place.

It is in the spirit of opposing armchair speculation that Anastasi opposes the interpretation of IQ tests as measures of general cognitive capacity. As to the content of the tests, she opposes the concept that the items used are examples of unlearned abilities, qualitatively different from achievement tests:

> The increasing efforts to prepare achievement tests that would measure the attainment of broad educational goals, as contrasted to the recall of minutiae, also made the content of achievement tests resemble more closely that of intelligence tests. Today the difference between these two types of tests is chiefly one of degree of specificity of content and extent to which the test presupposes a designated course of prior instruction.[21]

As school instruction gives even greater attention to general patterns of cognitive development, even this distinction is beginning to disappear. "Spontaneous" operations of thought have been recognized to be products of definite conditions of development and subject to educational measures. This is all the more true as these operations of thought become more complex and reach the level of modern scientific theories and methods. An arbitrary line between basic unlearned and general intellectual abilities, and derived, acquired and specific abilities cannot be drawn—at least not in the way supposed by IQ ideology.

As to the meaning of the predictive validity of IQ scores, Anastasi warns against explanations of this fact from basic differences in "capacity":

> Only in the sense that a present behavior sample can be used as an indication of other, future behavior can we speak of a test measuring "capacity." No psychological test can do more than measure behavior. Whether such behavior can serve as an effective index of other behavior can be determined only by empirical tryout.[22]

Thus, from this behaviorist viewpoint, speculation on the meaning of IQ scores appears unwarranted. A particular "behavior sample" simply correlates with other behavior. No inference can be drawn from observed behavior to unobserved "capacity." No assertions can be made about this behavior sample other than that it seems to be somewhat more general and less explicitly the object of school instruction than other "behavior samples" included in achievement tests. The behaviorist standpoint taken by Anastasi represents, on the one hand, a response to the mounting criticism against the pretensions of IQ tests. On the other hand, it represents an attempt to *preserve them* within an "agnostic" and positivist philosophical framework. Aware of the abuses of IQ tests, Anastasi attempts to remove the philosophical, "metaphysical" or speculative interpretations of IQ tests while preserving them on utilitarian grounds. After all they do correlate with other behavior, whatever the reason. Thus, on the use of IQ tests in the early days, she writes that:

> The application of such group intelligence tests far outran their technical improvement. That the tests were still crude instruments was often forgotten in the rush of gathering scores and drawing practical conclusions therefrom. When the tests failed to meet unwarranted expectations, skepticism and hostility toward all testing often resulted. Thus the testing boom of the twenties, based upon the indiscriminate use of tests, may have done as much to retard as to advance the progress of psychological testing.[23]

Between the materialistic and developmentalist explanation of intelligence and the idealistic interpretation of intelligence as a fixed inner capacity stand the intermediate philosophical position of agnosticism and empiricism, represented here by behaviorist defense of IQ tests without pretending to "speculate" on their meaning. Directed against the hereditarian interpretation of IQ, which is being defended by Jensen today, it is also directed against critics of IQ tests who argue that they are *inherently* loaded with unwarranted conceptions. On the one hand, the criticism of the fallacies of IQ ideology serves to "demythologize" the IQ test, at least in the minds of some specialists. But the agnostic position which consists in refraining from explanations, while simply noting associations of data or statistical correlations, leaves the door open to Jensen and others who express impatience with such "fastidiousness." The attempt to provide an explanation or to propose hypotheses is inherent in the process of scientific development. Jensen takes advantage of this fact in challenging both the agnostic refusal to put forward an explanation and environmentalist explanations generally. Moreover, since many psychologists use behaviorist concepts and methods, they are poorly prepared philosophically to defend the environmentalism they may in fact espouse.

The point of view which we are defending here is not one which seeks to "retard . . . the advance of psychological testing." By examining the presuppositions of IQ testing we are attempting to show that *this type* of psychological test has a very limited value which is basically far outweighed not only because of "unwarranted expectations," but because of features that are inherent in its essential make up.

# 6

## Differences: in the children or in the schools?

$O$ne of the principal features of IQ tests is the fact that differences in specific school performances can be correlated with differences in performances of a certain "general" character which are not normally the object of specific school instruction. We have seen that Galton attempted to "measure" differences in "intelligence," as judged by himself, with differences in performance on tests of sensory discrimination. "Measurement" here meant that the rank order of performance on the sensory discrimination test would correlate with rank order in intelligence as judged by (Galton's) common sense, and that the differences in these more "basic" and "general" abilities *explained* differences in intelligence as manifested in actual life by scholastic and social success.

### *"Fluid" and "crystallized" intelligence*

Jensen asserts that "thinking" is something that cannot be taught.[1] The child comes to the school with certain capacities, " ... an attention span long enough to encompass the teacher's utterances and demonstrations, the ability voluntarily to focus one's attention where it is called for, the ability to comprehend verbal utterances and to grasp relationships between things and their symbolic representations, the ability to inhibit large-muscle activity and engage in covert 'mental' activity, to repeat instruction to

**69**

oneself, to persist in a task until a self-determined standard is attained—in short the ability to engage in what might be called self-instrumental activities, without which group instruction alone remains ineffectual."[2]

These pre-school capacities are those which IQ tests are supposed to tap. They are themselves said to be essentially untaught. They differ in important ways between children, so as to explain their differences in school and social performance. So great are the differences that Jensen distinguishes between *two* qualitatively different forms of intelligence, "level one" and "level two." Level one involves associative rote type learning, whereas only level two involves higher level intellectual functions, normally associated with "really thinking," with "abstract intelligence." Jensen does not venture to define a precise point on the IQ scale where one can distinguish children with these fundamentally different capacities, but he argues that different forms of teaching should be evolved to conform to these different capacities. It would be cruel, Jensen writes, to impose unsuitable forms of instruction on children incapable of them, simply on the ideological grounds that all children are equal.[3] And Eysenck raises the problem of the compatibility of equality with liberty—opposing those "dogmatic equalitarians" who would force a single world of education on all children.[4]

Despite the fact that intelligence tests have been evolved atheoretically, and despite Jensen's admission that psychologists do not agree on a definition of intelligence, he nevertheless suggests that "the common feature of all such intercorrelated tests seems to be their requirement of some form of 'reasoning' on the part of the subject—some active, but usually covert, transformation of the 'input' (the problem) in order to arrive at the 'output' (the answer)."[5]

IQ tests do not normally measure differences in this ability *directly*. Jensen cites Raymond B. Cattell's distinction between "fluid" and "crystallized" intelligence.

*Fluid* intelligence is the capacity for new conceptual learning and problem solving, a general 'brightness' and adaptability, relatively independent of education and experience, which can be invested in the particular opportunities for learning encountered by the individual in accord with his motivations and interests.

*Crystallized* intelligence, in contrast, is a precipitate out of experience, consisting of acquired knowledge and developed intellectual skills.[6]

IQ tests normally measure both kinds of intelligence, i.e., basic intellectual capacity, and secondary "developed" skills. The two normally are "naturally correlated in a population sharing a common culture, because the acquisition of knowledge and skills in the first place depends upon fluid intelligence."[7]

In other words Jensen admits that IQ tests do not consist solely of items that "tap" basic "fluid" intelligence, but also uses items involving "developed" skills. How does one know that a particular item is a measure of "fluid" intelligence, while another measures the secondary "crystallized" variety?

## *"Culture-fair" tests*

Admitting that the Stanford-Binet does not clearly separate these two layers of intelligence, Jensen cites Cattell's Culture Fair Tests and Raven's Progressive Matrices as good examples of "Tests that measure mostly fluid intelligence" inasmuch as they "minimize cultural and scholastic content."[8] The fact that these tests "mostly" measure pure "fluid" intelligence and "minimize," but do not totally exclude cultural and scholastic content is an important concession, which an unalerted reader might easily overlook. The fact is that, as Anastasi admits, there have been *no* tests that are truly "culture fair," in the sense that they involve material and skills that are common to all cultures.

In fact, it is difficult to imagine a more "scholastic" test than Raven's Progressive Matrices which consists in finding designs that fit established and progressively more difficult patterns. The formal character of the test, its very emptiness of concrete content, makes it more of a test of facility in the abstract scholastic type of operation than the verbal tests of comprehension found in the Stanford-Binet. It does not seem likely that someone who was not already familiar with the types of formal literate skills involved in the school situation would do well on such a test. Anastasi writes simply "Studies in a number of non-European cultures, however, have raised doubts about the suitability of this test for groups with very dissimilar backgrounds."[9]

Another example of a supposedly "culture-fair" test is the Goodenough Draw-A-Man Test, in which a child is asked to draw the figure of a man. Points are given for inclusion of various parts of the body. There is age discrimination, as older children produce more complex drawings than younger children, and norms are established for each age. The test, moreover, correlates positively with the Stanford-Binet. One cross-cultural study using this test (standardized on the U.S. population) showed averages for different cultures, ranging from "IQs" of 124 for "suburban children in America and England, for children in a Japanese fishing village and for Hopi Indian children" to 52 for a nomadic tribe in Syria and 53 for a nomadic tribe in the Sudan. The authors do not turn to the concept of differences in "fluid" intelligence or "g" to explain these differences. As to the high scoring cultures, it is noted that "children grow up in continual contact with representative, graphic art." While at the other end of the spectrum, they point out that the "Muslim religion prohibits contact with graphic art. Yet, even among groups of Arab, Muslim children, the mean IQs for the Draw-a-Man Test range from 52, for the children of the Syrian Bedouins who had almost no contact with graphic art, to 94, for the children of Lebanese Arabs in Beirut who have repeated contact with the graphic art of Western civilization—even that including television."[10] Although originally F. L. Goodenough thought that his test had solved the evident cultural biases of the Stanford-Binet and came closer for measuring "pure" intelligence, such findings led him to conclude, not that the range of fluid intelligence among world cultures is enormous, but that "the search for a culture-free test, whether of intelligence, artistic ability, personal-social characteristics, or any other measurable trait is illusory."[11]

What items therefore "mostly" measure innate mental capacity? Eysenck picks an example of an item which, on the surface, seems to match the requirements of a "culture-fair" test when he distinguishes "Black is to white as high is to: green, tall, low, grey" from "Jupiter is to Mars as Zeus is to: Poseidon, Ares, Apollo, Hermes." The second is a measure of "achievement" while the first is supposed to reflect not special crystallized knowledge, but pure aptitude. Of course, Eysenck writes in reply to critics, there must be some elements of "culture"—such as knowledge of the English language. But given "equality of opportunity" here, differences in this type of item would measure unlearned aptitude to recognize opposites.

The example is highly deceptive. Eysenck does not tell us at what age level the item is located. If such a question was addressed to a five year old, it may not make any sense. The ability to recognize a relation of opposites is itself a developed ability. At a certain stage of development such a simple example might in fact discriminate between children who are "ahead" from those who are "behind." Eysenck seems to suggest that there are simply a lot of people with "low IQs" who are unable to answer the question correctly. The unqualified assertion that this is a good IQ item suggests to the layman that there must be a lot of people who are truly stupid, if they are unable to answer such a simple question. Of course, IQ tests composed of such items—as constructed for example by Eysenck—include far more complex examples of opposites, matrices, etc., and often include a time factor. Skill at solving abstract problems of this type, as well as the patience and motivation needed to grapple with them, may very well correlate with school success. This does not prove that different levels of mastery of such skills are unlearned, reflecting differences in basic "fluid" intelligence, nor that they are fundamental to and explain differences in more "developed" skills. The ability to recognize opposites might be acquired earlier than, for example, knowledge of geometry. No doubt, there is a developmental connection between such primitive intellectual skills and more advanced ones. But the classification of the first as a measure of "aptitude" or "capacity" and the second as a measure of "achievement" reflects the speculative constructions of IQ theory rather than a scientific theory of the real dynamics of the learning process.

One of the lines of argument which Jensen takes to support his biological thesis of the nature of intellectual differences relies heavily on this unscientific concept of "culture-fair" tests. Thus, in reply to the charge that the Stanford-Binet and other verbal IQ tests are biased against Black children, Jensen replies:

> Reputable tests, called "culture-fair" tests, do exist however. They use non-verbal, simple symbolic material common to a great many different cultures. Such tests measure the ability to generalize, to distinguish differences and similarities, to see relationships, and to solve problems. They test reasoning power rather than just specific bits of knowledge. Surprisingly, blacks tend to do better on the more culture-loaded or verbal kinds of tests than on the culture fair type.[12]

Jensen does not say that *all* cultures use the symbolic material involved in the test. Nor does he attempt to demonstrate that all segments or classes of a literate society which does require the formal logical use of symbols have equal opportunity to develop facility in such basically abstract mental operations.

The fact that not all cultures do score well on such tests is an important point that has profound implications for a scientific understanding of the development of "intelligence" understood as involving formal logical operations. Jensen passes by this point by asserting that "a great many different cultures" can be tested by such "culture-fair" tests. Anastasi refers to such non-verbal tests which are "valid" for European cultures, and which predict school success to some degree. But the absence of cross-cultural validity points out the important fact that the skills involved in such tests are historically developed.

### Literate intelligence: a product of history

The fact that one of the "cultural elements" admitted by Eysenck to be presupposed by his culture-fair item is literacy is another common sense observation, whose scientific implications are simply swept aside. For however natural the use of written symbols and simple logical operations may appear to us, they do not spring "naturally" from the human brain. An historical analysis of the development of literacy shows on the contrary that this was the result of fundamental changes in the productive forces of society and in its social organization, as well as a contributing force for further development. Such a phenomenon as literacy and the subsequent development of formal logical thought can be fully explained without hypothesizing some radical mutation in the genes. This is not to say that literate thought does not result in some form of reorganization of the human nervous system, and presupposes such a possibility in the brain—which has already evolved in relation to the development of artificial symbols, language, as a basic factor in the separation of the human species from the higher primates. But the realization of these possibilities depended on the evolution of human social organization rooted in changes in the productive forces *outside of* the human biological organism. And literacy involves also a transformation of the method of using symbols which takes place *outside of* the human brain.[13]

The Marxist historian George Thomson analyzes the social-historical context in which one form of literacy, the phonetic alphabet, was developed, and the profound consequences this had on further social, economic and intellectual development. The Phoenician alphabet evolved in the context of epochal changes in the mode of production in the Near East with the challenge of the growing force of commerce to the old regime:

The Phoenician alphabet of twenty-two letters, evolved to meet the needs of trade, was a great advance on the Mesopotamian and Egyptian scripts. These were so complicated and clumsy as to require the services of professional scribes, and the scribes were so well organized, under the protection of the state, that they resisted successfully the simplification of their art. The Greeks took over the Phoenician alphabet, probably in the ninth century B.C., and added signs for the vowels.

The invention of the alphabet may be compared with the invention of iron-working. Just as iron made metal implements more widely available so the new script made it technically possible for everybody to read and write. We have no means of determining the percentage of literacy in the Near East, but in the Greek democracies it must have included the great majority of the citizens. . . .

The alphabet embodied a new principle. The antecedent forms of writing—pictographic and ideographic—were concrete: the written symbol was a visual image of the idea it represented. It was not necessarily pictorial, any more than the spoken word was onomatopoeic, but it was concrete. Alphabetic writing, on the other hand, is devoid of visual imagery. The written word is a combination of symbols which are meaningless in themselves, being designed to represent the smallest phonetic elements to which the word can be reduced. In this way the new medium marked an advance in the development of abstract thinking, made it possible for speech and thought to become objects of cognition, and so prepared the way for the sciences of grammar and logic.[14]

We have quoted this passage at length because it shows clearly how intellectual "capacities" evolve in a definite socio-historical context. The methods of thought, which we today take for granted as given, are not innate, but depend on definite methods of thinking, definite techniques of thought that evolve on the basis of a reorganization of symbols external to the human biological substructure and have definite social presuppositions. The development of a science of language and thought presupposes that thought has been externalized in written form. The development of logical operations, moreover, involves further developments of methods of thought that became possible with the development of literacy. By disregarding or taking for granted the existence of these common, historically developed methods of thought, and solely focusing on *differences* in their execution, IQ tests, and the definition of intelligence implicit in IQ "operations," give the appearance of measuring something that is ahistorical. By abstracting from or taking for granted the historical evolution from preliterate to literate methods of thought, IQ scores appear to be ahistorical. But this appearance is a result of a methodological limitation of the test which considers past historical developments as "given." It is not a finding of scientific study of intelligence. A scientific study of "intelligence" would not only examine *differences* in the execution of a certain skill, but would attempt to explain the genesis of that skill itself. By investigating its causes, its historical origins, internal mechanisms and later developments, a scientific study would examine an historical, changing and developing phenomenon. The impression of changelessness given by IQ tests consists in simply ignoring change. The impression that a skill is basic, and biologically derived, consists in ignoring its prior conditions and the social-historical factors involved in its development. This abstraction from the historical character of the content and skills tested by IQ tests is facilitated by the choice of kinds of skills which are relatively primative—such as the ability to recognize formal opposition.

What of the enormous advances in intellectual skills and knowledge which have taken place since the time of Plato and Aristotle? Should these be relegated to the sphere of "crystallized" knowledge while the petrified forms of ancient logic are given the title of "fluid intelligence?" Does the first represent "mere achievement," "bits of knowledge," environmental "material," while the latter engages "capacity," reasoning power and just plain thinking? In fact, in any area of real knowledge, the main condition for "intelligence" in that area is the presence or absence of a scientifically validated theory which explains the basic laws and events occurring in that particular field. Next to this, whatever "brightness" or "quickness" an individual may show in the exercise of formal logical thought may appear to be the darkest dull-mindedness. It is in this sense that Engels attributed to the discovery of certain basic concepts in anthropology the main characteristics of "intelligence":

This rediscovery of the primitive matriarchal gens as the earlier stage of the patriarchal gens of civilized peoples has the same importance for anthropology as Darwin's theory of evolution has for biology and Marx's theory of surplus value for political economy. It enabled Morgan to outline for the first time a history of the family in which for the present, so far as the material now available permits, at least the classic stages of development in their main outlines are now determined. That this opens a new epoch in the treatment of primitive history must be clear to everyone. The matriarchal gens has become the pivot on which the whole science turns; since its discovery we know where to look and what to look for in our research, and how to arrange the results. And, consequently since Morgan's book, progress in this field has been made at a far more rapid speed.[15]

No doubt the ability to recognize formal opposition, etc., continues to play a role in scientific development. But did Morgan's theory evolve out of a superior capacity to recognize opposites in his field? In fact, as Engels points out, Morgan made his advances by breaking past the barriers erected by the rigid opposition thought to exist between exogamous and endogamous clans, which Engels attributes in part to the formal "legalistic mind" prevalent in English scientific circles. In any case, it is the existence of definite, scientific concepts and general ideological framework that orient the development of a science and determine by and large the rate of progress that will be made. In any real field of knowledge it is the presence or absence of pivotal concepts that make the *decisive differences*. Such differences can only be measured by taking the long view of scientific development. At any one point, of course, one will find that individuals may differ in the exercise of such concepts. But should we limit our definition of intelligence to these relatively *minor* differences in given skills or concepts?

By focusing solely on the differences that exist at any given moment, and by "measuring" these differences with items involving the most limited forms of intellectual activity which have remained unchanged for 2,500 years, IQ theory pretends to touch on something ahistorical and so "hypothetically" biological. What it in fact touches on is only the ahistorical character of its own methods and the biological hypothesis which, far from attempting to validate by looking at the real phenomena of intellectual history, it only tries to pass off by a sleight-of-hand.

From an historical point of view, we see instead enormous advances in real intellectual and practical capacities rooted in modern industry and science. Such real capacities cannot be found by scrutinizing the "naked ape" stripped of all the instruments of knowledge and of practice that make human beings to be truly human. Such differences in historical development become evident once one enlarges one's perspective beyond the "given" to look at the changes that have occurred in our tools of thought and action. It was such a perspective that led Engels in the 1870s to remark that "We have . . . infinitely multiplied production, so that a child now produces more than a hundred adults previously did."[16]

### Thinking can be taught

Essential to the recognition of the historically developmental character of "intelligence" is an understanding of how knowledge grows, and a science of learning and teaching. Thinking can in fact be taught. In the past, thinking in the sense of formal reasoning was taught more or less by a spontaneously developed and unconscious method. This led to the belief that thinking either happened or it didn't happen. Progressive educators today recognize that intellectual development is not a fatal process in which the basic assimilative faculties are more or less fixed at least by the age of six or seven. Pedagogical advances make it more and more possible to facilitate the development of basic intellectual skills in everyone. This is the ideal of equality that alone makes sense. It is not a question of making everyone equal in all departments of knowledge and practical skill—an impossibility given the necessity of specialization. But specialization and intellectual division of labor presupposes a whole and interaction between the parts of the whole. This in turn depends on general understanding of at least the fundamental concepts and

methods in the various areas, and philosophical generalization of
the concepts and methods common to the whole picture. Is it too
much to ask that modern education aim at providing all children
with such fundamental theoretical and practical knowledge?

Are there children, not seriously handicapped, who are incapable of being intelligent in this sense? Pedagogical advances make it
more and more possible to facilitate the development of basic intellectual skills in everyone, even after it has become "too late" according to some versions of the nature of intellectual development. Ruth
Beard cites Soviet concepts of pedagogical technique, resting on the
general dialectical materialist view of intellectual development:

> The possibility of deliberately speeding, or aiding, the process of
> internalization has been investigated in Russia. Galparin and
> Talyzina (1961) used the same method both with normal children
> of six and with backward adolescents of fifteen and sixteen. The
> children initially engaged in activities with concrete materials,
> then progressed to making audible descriptions and instructions,
> concrete aids being gradually reduced until, finally, the concepts
> were interiorized in verbal form. With this approach concepts in
> elementary geometry which had previously proved impossible to
> the backward adolescents were formed almost faultlessly from
> the beginning. The physical responses needed to be fully de-
> veloped initially and every step in forming the concept had to be
> worked through; but in learning subsequent related concepts
> some steps could be omitted and the students might begin at the
> stage of thinking out loud or even at the purely mental level.[17]

Such an approach to education does not rest on "egalitarian
dogma" but on an attempt to grasp the basic laws of intellectual
development so as to facilitate the process of education. Such an
approach is characteristic of progressive educational theory in the
United States, Great Britain, France and elsewhere where the IQ
theorists seem to be fighting a rear guard, if still important effort to
preserve the theory of innate differences in intellectual capacity.
This theory in turn spills over into the social realm where society is
made out to be based on a natural hierarchy between people. While
Eysenck and Herrnstein make much of attacking "dogmatic
egalitarianism" and socialist utopias, Brian Simon points out that
Soviet educational psychologists do not stifle differences between
children or turn away from talent.[18] Nor is the recognition of spe-

cially talented children a contradiction of "socialist dogma." It is a matter of fundamentally different approaches to pedagogical techniques. Psychometrics is concerned with sorting out and selecting children so as to maximize individual differences and is bound up with a tracking system that builds in and intensifies differences. Soviet education rests on essentially unstreamed and group-centered education which involves individual children in the educational development of their peers. Recognition of especially talented or backward children in this system nevertheless goes on. Simon points out the difference in approach by educational science to these differences. It is not a matter of classification on the basis of a theory of unchangeable properties, but of understanding the mechanisms of learning more thoroughly. The "gifted" child in this context is not only involved more directly in the development of his or her peers, but becomes a subject of scientific study aimed at a deeper understanding of the methods of learning (just as the "gifted" teacher provides a subject for research aimed at generalizing from the actual approaches used by such an outstanding person).

To conclude, we have raised historical and philosophical objections to one of the main concepts of IQ theory—that differences in "performance" can be explained by differences in "capacity," and that the latter can be more or less directly measured by performances on certain types of IQ test items involving certain kinds of "abstract reasoning" skills. We have argued that such skills do not indicate some basic "fluid" intelligence which can explain later "crystallized" intelligence. This is not to deny *correlation* between performance on the tests of formal "abstract reasoning" and performances in school subjects in general. Nor are we denying the possibility of certain causal connections between deficiencies or proficiencies in certain primitive and fundamental concepts, and cognitive skills and success or failure in school. In this sense remedial work may have to go back to "basic" skills to recreate the foundations of cognitive development. But the extent that IQ test items may actually provide clinically useful information for such an analysis of the intellectual substructure of an individual is essentially the result of chance and guesswork, rather than of a scientific study of the most general features of intellectual development. In any case, the rule of thumb by which "content validity" is attributed to IQ tests is simply a common sense assumption that certain kinds of items are not learned—and so plausibly tap basic potential, etc. We have seen that this concept does not hold up under empirical and theoretical analysis.

## Inequality of opportunity

Far from upholding "dogmatic equalitarianism" Marxist theory in fact places the main inequalities in the *social* conditions of intellectual development at the heart of its theory of history. The passage cited from Thomson earlier underlines the social conditions of the revolution in intellectual method that constituted the creation of alphabetical writing. The conditions of development made possible by new productive forces, especially with the use of iron, and the development of extensive trading in the Mediterranean area, rendered the old pictographic script obsolete. An elite body of scribes who had the privileges and leisure to devote themselves to the specialization necessary to master a complex art stood in the way of the diffusion of literary skills. Employment of these specialists was expensive, and complicated the processes of exchange which trade necessitates. The development of the alphabet enormously simplified the task of literary representation and made it possible for a much broader sector of the population to achieve literacy. Did such a "simplification" of the methods of representing language result in a decline in "intelligence?" In fact, as Thomson makes clear, the more democratic extension of literary skills made possible a qualitative development in science, art and culture.

It should not be forgotten, however, that the extension of literary skills to the free population of Greece took place under conditions of slavery, where the majority of the population was subject to the most brutalizing forms of manual labor. So entrenched was this division of intellectual and manual labor that Aristotle, reflecting common opinion at the time, ranked the arts and crafts by the extent to which they were independent of physical labor (with sculpture, for example, occupying a low rank). At the one end of the social spectrum, those who were totally subject to physical labor, the slaves, were not regarded by Aristotle as truly human beings. At the other end of the scale were those moments when an individual, involved in pure intellectual thought, came closest to the life of the gods.

The illusion that thought is a pure activity separated from and guiding practice "from above" stems from social-historical conditions which led to the division of intellectual and manual labor in societies based on private ownership of the material and mental means of production. Marx wrote in the *German Ideology* that

> The ideas of the ruling class are in every epoch the ruling ideas: i.e., the class which is the ruling *material* force of society, is at the same time its ruling *intellectual* force. The class which has the means of material production at its disposal, has control at the same time over the means of mental production, so that thereby, generally speaking, the ideas of those who lack the means of mental production are subject to it.[19]

The specialists in intellectual activity, Marx wrote, are in general dependent on the dominant "practical class," so that, ideologically active, the intellectuals in general are passive in relation to the practically active ruling class which demands an ideology to justify its class rule as well as a level of scientific knowledge commensurable with the needs of social production. And as a whole the working people who are exploited in any social system are deprived of the advanced means of mental production, and subject to the dominant ideology. This does not prevent the development of forms of popular culture and "intelligence" which reflect the concrete experiences, aspirations and struggles of the oppressed sections of the population. In times of social revolution, new social forces tend to express their interests in ideologies which more or less clearly challenge the traditional ideologies and in this way prepare the intellectual climate for social progress.

This brief sketch of the Marxist theory of intellectual development requires further development for application to today's conditions of life in which for the first time in history the material forces of production make it possible to look for an end to the age-old division of intellectual and manual labor. Marx regarded this as possible only when social production has been controlled by society as a whole, and when the level of production is such that it will be possible to organize society on the basis of the maxim "from each according to his ability to each according to his need."

The division of intellectual and manual labor continues to be a major structural feature of capitalist (and in a different form of socialist) society. Consequently, it should not at all be surprising to a Marxist to find that children from business and professional fami-

lies should do better on tests measuring scholastic abilities than
children from working class families, and that among working class
families the children of skilled workers should in general do better
than those of unskilled workers. Finally, it is not surprising, although it is still appalling, that children from racially and nationally oppressed peoples, those suffering from what Marx called
super-exploitation, involving the highest levels of unemployment,
should receive the worst forms of education and achieve the lowest
levels of performance on tests geared to measure rank order in
school.

## IQ, tracking and segregation

Are IQ tests biased against working people and especially
Blacks? Our answer is unequivocal. They are. This is not so only
because the language is "white" and items are middle class oriented.
IQ tests are an obstacle to the educational advancement of working
people in general and Blacks in particular because they ascribe the
causes of differences in educational performance to internal deficiencies in capacity. And based on this fraudulent claim, which is
implicit in the structure and content of the test however much some
psychologists may back off from asserting it, IQ tests partially
result in and basically justify an educational system that builds in
inequalities through tracking within schools and classrooms as well
as between schools. This tracking system is taken for granted as a
fact of life in the overwhelming majority of schools. It is a fact of life,
so far as most people are concerned, in the qualitative differences
*between* schools. And most importantly, especially for the thesis
defended by Jensen, it is evident in the widespread racial segregation of schools in the U.S. This segregation is giving way, inch by
inch, and in the face of strong racist opposition abetted by politi-
cians in many cities and in the Federal government. The revival of
the "IQ Argument" is directed against this goal of educational
equality and integration of education of Black and white. Moreover,
at a time of serious economic crisis, the fortification of racism with a
variety of specious theories turns whites against the legitimate
grievances of Blacks, and away from the fact that they (the overwhelming majority of whites) also suffer from the ills of an exploita-
tive social system.

The main bias of the IQ test rests with its fundamental features and assumptions rather than with the particular, secondary biases of test items. A test constructed in the atheoretical common sense manner of the Stanford-Binet is bound to include items of moral opinion which reflect class biases, vocabulary reflecting racial and class bias, as well as items for which such types of particular bias have been eliminated. "Pure" IQ tests can then be constructed which eliminate all such particular bias, and nevertheless correlate (imperfectly) with the "biased" tests as well as with school performance. Such tests will be inherently biased because they imply, if they do not overtly state, that differences in school performance are due to differences in "capacity." The fact that the Stanford-Binet contains both, items reflecting ideological opinion and items involving general scholastic ability, is a result of the fact that those who do well in school tend to reflect both "middle class" ideological values and advantages for the formulation of formal intellectual skills.

Our opposition to IQ tests is therefore not on the grounds that such tests are wholly arbitrary and subjective, and that they simply measure working people on the basis of middle class biases.

We have been arguing that it is not surprising that tests of intellectual performance show intellectual inequalities between class and racial groups in a class society which reproduces a division of labor between mental and manual work. At the same time changing historical conditions necessitate and stimulate a change in this situation, a change reflected in greater self-confidence by the mass of the people. Contradictions in the economic, social, political and cultural development in society increasingly call into question the "capacity" of the ruling elite to manage society. As a result more and more people are dissatisfied with keeping their "place" in society, and are calling into question the foundations of the class system of social organization.

To the democratic aspirations of the people for a better life, as expressed in the desire for better education and expanding educational horizons, Jensen in effect replies, let's face it, occupations are inherently unequal—you yourself recognize it. But we can at least provide "equality of opportunity" if not of results. We can offer you a school system which recognizes neither class nor color. And if that isn't enough we have IQ tests which can overlook differences in "achievement," which an intractably unequal environment might

still produce, in order to perceive your native capacity. The perspective here seems little different from that of Plato whose Republic
would be based on allotting positions in society, not on the basis of
social privilege, but on the more "philosophical" recognition of inner
differences in the souls of men, making some fit to rule, some to fight
and others to work.

The crucial question is therefore whether the school systems do
in fact select children on the basis of true merit, or whether they
essentially reproduce the class differences and racial discrimination
which prevail in society at large. This is not a question to which this
book can devote any special attention. We simply wish to point out
and to question generally admitted features of the school system:
great, regional disparities based on the different methods of funding
schools; differences between urban and suburban systems reflecting
urban class and racial differences, and with access to different
sources of funding as well as to different clientele; racial segregation
of schools throughout most parts of the country; and finally selection and differential education of "fast" and "slow" learners from
the earliest years, "tracking" children within classes as well as
among classes. In no way is there the least plausibility to the notion
that children are provided with equal educational treatment, and
that despite all efforts directed to such an end, children simply
perform differently. When children from the first grade if not sooner
become aware that they are good readers or poor readers, are segregated in classrooms and given differential treatment, publicly rewarded or punished for differences which may be slight in the
beginning but which become hardened over time—then where is the
equality of opportunity?

The circularity of Jensen's whole argument is evident in this
spurious assertion: 1) We have tried hard to make all children equal,
based on a misguided concept of the "average child," 2) our efforts
have failed—as seen by IQ tests; 3) so let us now turn our attention to
providing different forms of education for different kinds of children. In fact, the different forms of education, the educational
inequality of the school system, has always been its most glaring
feature despite words to the contrary and the precarious and minimal efforts to overcome this inequality.

# 7

## *Dialectical relation of biology and society*

As if his arguments were not sufficiently convincing Jensen has frequently resorted, in various ways, to arguments from analogy with physical properties whose distribution is known to be governed by genetic laws. Thus in "The Differences are Real," in the popular monthly magazine *Psychology Today,* Jensen argues that the "genetic hypothesis" which "has not yet been put to any direct tests by the standard techniques of genetic research" must nevertheless be

> seriously considered ... for two reasons: 1) because the default of the environmentalist theory, which has failed in many of its important predictions, increases the probability of the genetic theory; 2) since genetically conditioned physical characteristics differ markedly between racial groups, there is a strong *a priori* likelihood that genetically conditioned behavioral or mental characteristics will also differ. Since intelligence and other mental abilities depend upon the physiological structure of the brain, and since the brain, like other organs, is subject to genetic influence, how can anyone disregard the obvious probability of genetic influence on intelligence?[1]

The first reason has already been partly answered in preceding sections where we have challenged the validity of a supposedly empirical method of examining intelligence which neglects its historical development as well as the social conditions of the distribution of the "means of mental production." Truly sociological and historical investigations can explain the development of scientific thought as well as the division of mental and manual labor without recourse to genetic hypotheses. Secondary problems of explanation require more concrete studies. Thus Jensen asks why standard socio-economic measures do not explain the average superior performance on IQ tests of Native Americans to Black Americans, since these measures show Native Americans to be even more impoverished than Blacks. A socio-historical approach to the study of Native Americans and Blacks would hardly be sufficient if it relied solely on the "environmental" indicators used by Jensen. Despite the oppression to which both peoples have been subject for hundreds of years, it is obvious that the forms of the oppression as well as the responses to it have been quite different. Centuries of slavery and centuries of genocidal wars and subjection to existence on impoverished reservations are two different forms of oppression which cannot be "measured" solely in terms of family income, parental education, housing, etc. Jensen writes that "No testable environmental hypothesis has yet been offered to account for these findings."[2] We hope that no such tests, involving the subjection of more peoples to slavery, will take place. This is not to say that an empirical study of past and present social conditions affecting educational differences could not explain any average differences in performance.

Jensen's second reason for a "genetic hypothesis" raises basic questions regarding the nature of intelligence which were implicit in our previous analysis of the socio-historical character of the intellectual skills which Jensen and Eysenck consider to be more or less natural. In this argument we have stressed the reductionistic and metaphysical character of IQ methodology which attempts to define intelligence in relation to relatively static formal-logical skills that have been in existence for over 2,500 years. Here Jensen again attempts to reduce intelligence to the level of the "natural" functioning of the brain. In arguing that developments in intelligence have occurred historically, without requiring any genetic hypothesis of qualitative transformations in the biological composition of the brain, we have presupposed that there is more to intelligence than spontaneous brain activity.

In studying the relation between biological and social factors of human development Marxist theory underlines the qualitatively new features that differentiate human evolution from previous biological evolution. To understand this qualitative development it is necessary to examine the main features of biological evolution. One must then compare these with essentially new factors which emerged in the evolutionary process which resulted in the appearance of laws of human development that are qualitatively different from those which operate in the evolution and functioning of prehuman species. While Darwin's theory of evolution explained the main factors of biological evolution—supplemented by Mendel's theory of genetics and later development of genetic theory—nineteenth century theorists attempted to apply biological laws to the explanation of human evolution. This "Social Darwinism," which developed in fundamental opposition to the theory of historical materialism, provided the theoretical foundations of Galton's search for an empirical test of the supposed biological basis of the differences between social classes. Social Darwinism (or perhaps "Social Mendelism") continues to underlie the "meritocracy" theory of Jensen and his followers. Reaffirming the main philosophical and historical principles underlying IQ theory Jensen cites the thesis of E. L. Thorndike in 1905 that "In the actual race of life, which is not to get ahead, but to get ahead of somebody, the chief determining factor is heredity." Jensen comments: "Since then, the preponderance of evidence has proved him right, certainly as concerns those aspects of life in which intelligence plays an important part."[3]

## *Darwin and Social Darwinism*

While stressing up to this point certain basic characteristics of IQ theory and its presuppositions, we have to turn to more specific scientific theories about biological and social evolution. Thus, in order to understand Social Darwinism and its contemporary versions, it is necessary to provide a brief outline of Darwinian theory of natural evolution. Darwin made a fundamental breakthrough in the knowledge of nature by discovering the main causes of biological evolution. In so doing he struck a decisive blow against the metaphysical, ahistorical concept of nature as consisting of eternally existing species. According to Darwin, nature does not consist in a collection of species co-existing from the time of creation, but of a dynamic process leading from lower to higher forms of life. Species do not simply co-exist; they interact with each other and the rest of their environment. In this interaction there are innumerable individual variations in any particular species. In particular, there is a prodigious number of potential beings, of the seeds of life, of which only a relatively small number are capable of reaching full maturity in a limited environment. In this "struggle for existence" those individuals who are better able to survive have a greater chance of living and reproducing. "Chance" variation provides the variability of the species, while nature "selects" those individuals most suited to the requirements of the environment. Darwin believed that, in general, changes in species took place gradually, until a point of evolution was reached in which the differences between individuals was so great that interbreeding ceased, and a distinctively new species emerged.

Natural evolution took place, therefore, because of a combination of "chance variation" and "natural selection." An overall progressive movement characterizes this process primarily because new variations build on early forms of organization, and further complicate the species. Higher forms of organization are selected by their adaptability to the highly complex features of an ever changing environment.

In this process, the role of the environment in selecting individuals for survival is decisive. Nevertheless, individuals must themselves "vary," and the infinite diversity of nature is an essential factor in its lawful progress from lower to higher forms. Darwin attributed this variation to "chance." Today we know that there are definite causes to genetic changes, and we know something about how they operate. The pioneer in an understanding of the simplest laws of genetic variation due to the combination of certain genes from the parent organisms was Gregor Mendel. Jensen in fact calls the revolution in thought which he advocates a "Mendelian revolution" because of Mendel's role in the founding of genetics.

In his own analysis of Darwin's theory, Engels, probably without knowledge of Mendel's breeding experiments, questioned the "chance" character of variation:

> It is true that Darwin, when considering natural selection, leaves out of account the causes which have produced the variations in separate individuals, and deals in the first place with the way in which such individual variations gradually become the charac-

teristics of a race, variety or species. To Darwin it was of less immediate importance to discover these causes—which up to the present are in part absolutely unknown, and in part can only be stated in quite general terms—than to establish a rational form according to which their effects are preserved and acquired permanent significance. It is true that in doing this Darwin attributed to his discovery too wide a field of action, made it the sole agent in the alteration of species and neglected the causes of the repeated individual variations, concentrating rather on the form in which these variations become general; but this is a mistake which he shares in common with most other people who make any real advance.[4]

The Mendelian revolution does not negate the main features of Darwinism, but rather specifies more precisely what Engels called its "field of action." To attribute full or dominant causality to genetic laws, and to underrate the role of environment, is to commit the inverse error. In his assault against "environmentalists" Jensen exaggerates the field of action of genetics. As a "Social Darwinist" (or rather a "Social Mendelian") he applies the laws of biological evolution to human evolution and overextends the field of action of these laws. But even in the domain of biology proper, Jensen's concept of "heritability" involves a failure to grasp the real relation of genes and environment in the biological process. We will return to this question later when we examine Jensen's argument for the high "heritability" of IQ.

## Biology and human evolution

Against Social Darwinist theory, the Marxist theory of social evolution points to a qualitative difference between the laws that operate in the evolution of prehuman species and those that reflect the new features which distinguish the human species from its predecessors. Essential to the emergence of *homo sapiens* was the development of social labor as the decisive factor in its survival and development. In his essay on "The Role of Labor in the Transition from Ape to Man" Engels described, on the basis of the evidence available in his day, the probable path that biological evolution took with the growing importance of tools to the survival and development of certain higher primates. At the same time, in the course of development of primitive forms of labor, articulate speech too grew in importance. Biological changes of fundamental magnitude, including especially the evolution of a more highly developed central nervous system, occurred as a result of the increasing importance of social labor, involving the use of tools and the development of language.

Thus, a biologically superior species emerged as a response not so much to the natural environment in its relatively pure and independent development, but to the use of tools and language which mediated the relation between the organism and its environment. Adaptation of human physiology to the use of tools and to the use of an artificial form of communication made the emerging human species *"self-developing"* in a qualitatively superior form than is the case with animals which were adapted to an almost wholly external, independently evolving environment. The human physiology emerged, not in a relatively passive response to changes in the environment, but in the more active process of changing the environment, using natural objects and transforming them, to attain ends beyond themselves. It is here that we begin to see the inadequacy for the explanation of human development of a strictly biological model of evolution, in which biological variation interacts with an external selective environment. The "environment" which was decisive in the selection of the higher biological form of the human species is already a "humanized" environment—naturally, still resting on the relatively external, partially uncontrolled natural environment and always rooted in natural laws.

We have said that for Engels "social labor" played a major role in the evolution of the human physiology and higher nervous system. In the early phase of transition from higher anthropoids to humanity, biological changes continue to be the main form of development. But with the emergence of a physiology and a nervous system more completely adapted to social labor, this latter is able to develop more complex forms as well. And in discussing the evolution of the forms of social labor we are no longer confined to strictly biological laws. The laws governing the evolution of tools from the simplest stone weapons to more advanced equipment depends on 1) structural characteristics of natural objects, and the potentialities for their development and 2) the development of human knowledge and ability, building on the knowledge and skills of past generations. Interrelated with this basis of distinctly human evolution, which Marx called the productive forces, are the forms of social organization which are in themselves a powerful productive force and indeed complemented the paltriness of early man's technical abilities. New *social* forces begin to operate which, while having as a necessary starting point the physiological grounds for speech and thought, nevertheless evolve according to new laws superseding, submerging or transforming biological and physiological laws. Just as the passage from stone to bronze to iron tools involves a logic of evolution which is inherent in the properties of these materials and requires the development of skills and techniques needed to make use of them, so the related transformation, development and changes in social organization depends on properties of social relations and not directly on biological or physiological properties. No doubt the biological features of sex and age play an important role in the organization of early human societies. But their structural relations and dynamics become increasingly due to laws which are not strictly reducible to biological or physiological laws.

Once a certain threshold of biological development is reached, the changes that become ever more decisive for determining the mode of behavior of human beings are changes in the productive forces, the social organization of society, and the body of knowledge and values passed on by cultural mechanisms built up over time. Such processes are no doubt accommodated to the features of the human organism. On the other hand, more essentially, it is the human organism and its greater "plasticity" that is essentially accommodated to human labor.[5]

Marx succinctly distinguished between animal life and human life when he wrote in 1844 that "The animal is one with its life activity. It does not distinguish itself from it. It is *its life activity*. Man makes his life activity itself the object of his will and of his consciousness. He has conscious life activity. It is not a determination with which he directly merges."[6]

Marx distinguishes human universality from the "particularity" of animal species which are "one with their life activity," i.e., which exist primarily according to the laws of their own organic make up. Man potentially makes all of nature his "inorganic body," and confronts his own life activity as an object of his activity. Although man's organic existence remains relatively stable, changes in this external component of human existence can proceed indefinitely, so that the whole of nature can potentially serve human existence. While each animal species has its particular "ecological niche," the human body is adapted to social labor which continues to develop and to transform any particular environmental limitation.

Thus human thought, Marx writes, is basically objective and capable of unlimited development. It is not the organic needs of animal life that determine the human outlook on life. Human thought is mediated by language, an artificially developed organization of sounds and other symbols which permit the human brain to represent more than what is present in immediate sensation, direct forms of memory and biological instinct. Because it is mediated by language and externalized in human discourse (and eventually in writing), human thought is an "object" for human activity. It can be shaped and perfected over time. Contrasting experiences can be confronted with one another, present forms compared with past forms. Human thought is therefore not limited to one aspect of nature, one side of the environment as it directly impinges on immediate, organic need. Human thought is capable of grasping the many-sided, overall development of nature. While an animal one-sidedly "understands" those features of the environment that "interest it" according to its species needs, human thought, as a result of language and social exchange, is able to grasp all the sides of a process and ultimately discover its basic laws.

Jensen appeals to "common sense" when he asks:

> How can a socially defined attribute such as intelligence be said to be inherited? Or something that is so obviously acquired from the social environment as vocabulary? Strictly speaking, of course, only genes are inherited. But the brain mechanisms which are involved in learning are genetically conditioned just as are other structures of the organism. What the organism is capable of learning from the environment and its rate of learning thus have a biological basis. Individuals differ markedly in the amount, rate and kinds of learning they evince even given equal opportunities. Consider the differences that show up when a Mozart and the average run of children are given music lessons![7]

The statement that "capability of learning from the environment" has a "biological basis" is an example of philosophical ambiguity or duplicity.

"Capability of learning from the environment" is an expression that reflects the physiological theory of learning applicable to the higher species of animals. It fails to distinguish the qualitatively new *form* of learning that became possible with the development of language as a socially created instrument of thought. In the classical language of Pavlovian theory which has been developed and refined by contemporary Marxist psychologists, there is a failure to distinguish between the "primary signal system" of the higher animal species from the "second signal system" that develops with the human species.[8] The latter depends on the former, is conditioned by it, but introduces qualitatively new features into the learning process. As a result, human thought is not restricted in its learning capacity by the limits of its sensory organs and by the relatively limited practical needs of an organism that has evolved in relation to a relatively specific environment. Because human learning takes place through the instrumentality of language its capacity for growth in *unlimited*. This is true not only in the quantitative sense that knowledge can be stored up to an extent limited only by the techniques of organizing knowledge—techniques which have evolved from the preliterate, oral culture of primitive peoples, through literate modes of transmitting knowledge, to the computer-age leap in storing and coordinating knowledge. But together with this quantitative side, there is a qualitative progression of human knowledge toward greater scientific objectivity. Here theoretical advances play the decisive role in the "rate of learning" that is possible today. The greatest "brains" in the middle ages were groping in the dark compared with the rapidity of learning that is possible today as a result of vast developments of scientific theory and techniques of processing knowledge, both of which involve the cooperative effort of thousands and even millions of humans.

For human beings an external environment does not select already present "internal" capacities; the environment has selected a species whose own capacities are externalized and evolve on the basis of changes in their external forms of existence. Thus, contrary to biologistic theories of the development of human capacities, Marx wrote pointedly (and humanistically) that "music alone awakens in man the sense of music . . . Only through the objectively unfolded richness of man's essential being is the richness of subjective *human* sensibility (a musical ear, an eye for beauty of form—in short *senses* capable of human gratification, senses affirming themselves as essential powers of *man*) either cultivated or brought into being."⁹ On the other hand, stripped of the objective conditions of *human* development, human beings can become brutalized:

> For the starving man, it is not the human form of food that exists, but only its abstract being as food. It could just as well be there in its crudest form, and it would be impossible to say wherein this feeding activity differs from that of *animals*. The care-burdened man in need has no sense for the finest play; the dealer in minerals sees only the commercial value but not the beauty and the unique nature of the mineral; he has no mineralogical sense.¹⁰

### Biological changes in human evolution

Biological evolution does not stop with the appearance of social labor. Natural selection continues to play a role, particularly in the early development of humanity—when the forms of social labor were still relatively primitive. Thus, in his study of the early forms of the family, Engels appears to juxtapose the process of development of social labor with the process of natural selection. He argues that the succession of the early forms of the family, in primitive hunting and fishing and early agricultural societies, was due to natural selection. Engels hypothesizes that possible marriage partners were successively narrowed down to exclude, first marriage or sex between parent and child, then marriage between brother and sister, and finally marriage within the clan. "Exogamous" marriage became the norm according to which an individual could only marry outside his or her clan. The resulting "pairing family" evolved, Engels writes, because of natural selection—i.e., presumably because the offspring of such marriages were healthier and less subject to the ill effects of close inbreeding. The further course of evolution of the family, particularly the evolution of the patriarchal family and the appearance of the monogamous family based on male supremacy, developed because of new *social* forces. "Natural selection, with its progressive exclusions from the marriage community, had accomplished its task, there was nothing more for it to do in this direction. Unless new, *social* forces came into play, there was no reason why a new form of the family should arise from the single pair."¹¹

Whatever the validity of this particular analysis, there is for

Engels a continuing but declining role of biological evolution, during the early period of humanity when the social productive forces, although decisive in the essential physiological constitution of the species, remain as yet relatively undeveloped. This concept is rooted in the dialectical theory of development and transition from lower to higher forms of existence. During a transitional phase the two types of laws appear to be juxtaposed until the one is clearly subordinated to the other. This is not to say that preceding laws cease to function, but rather that they cease to explain the dynamics of change and development of the higher form of existence.

The precise interrelation of two sets of laws—e.g., chemical and biological—must be made the object of a special study. But such a study must be guided by a general overall theory of how the two sets of laws interact. It is characteristic of metaphysical thinking to isolate the features of each phenomenon and to juxtapose them. Such abstract juxtaposition may be possible in thought. But no such thought model should be forced on reality to constrain it to fit such a mold. Yet this is what is done when "heredity" and "environment" are examined abstractly and one attempts to define how much of each contributes to the phenomenon. Biological laws of inheritance, together with natural selection, are not simply juxtaposed to "environmental"—i.e., social factors to explain a particular phenomenon. We will return to this subject when we come to "heritability" analyses, which purport to establish the quantitative contribution of heredity and environment to the phenomenon (and not to the phenomenon in itself but to its *variations* within a group, a concept that above all calls for some critical evaluation).

## Race

The transitional, but declining, role of natural selection in the development of *human* organisms—i.e., organisms *primarily* adapted to social labor—is of crucial importance in understanding the formation of races. Soviet physical anthropologist M. Nesturkh summarizes basic Marxist concepts in his book *The Races of Mankind:*

As people spread over the face of the earth, they came up against differing natural conditions. Although natural conditions have such a tremendous effect on the species and subspecies of animals, they could not act so intensively on the races of man since humans differed qualitatively from animals in that they were constantly opposing themselves to the nature that surrounded them and transforming it in the process of collective work. There is no doubt that in the course of man's evolution many racial features possessed an adaptive character which was to a large extent lost as the role of social factors increased and that of natural selection gradually lessened and then disappeared . . . Migration, isolation, increase in numbers, the mixing of anthropological types, and change of food habits were, together with natural selection, the main factors in the process of race formation among the ancient hominids. Appearing in numerous combinations and differing in their intensity, they conditioned the differentiation of races, forming a network of anthropological types, at first sparse but later much denser, that were connected in varying degrees by transitional groups.[12]

Because of this complex historical past involving periods of isolation and intermingling, there are no "pure races." Many biologists distinguish races, not by absolute characteristics, but by the statistically greater frequency of biological characteristics such as blood types that distinguish one group relatively from another. From a more basic point of view, it is essential to underline the fact that racial formation is distinct from the formation of species and subspecies among animals, precisely because human beings are not "one with their life activity" but confront that life activity as an object that is subject to continual development in the relative opposition between man and nature. "Although adverse climatic conditions and natural barriers (high mountains, extensive dense forests, waterless deserts) hampered man's migration, they did not prevent it. Social organization, labor, clothing, tools, weapons, fire and means of transport served to counteract those natural factors that usually have differentiating effect on any species of animals."[13]

As a form of Social Darwinism, Jensenism continues to maintain that the decisive factors for change are "selection," by a relatively external environment, and biological variation of the organism. Modern "industrial society" is said to provide a social environment which selects for survival and success individuals who have the biological capacity for operating successfully. Superior "IQ genes" are said to give certain individuals and races this advantage. In opposition to this biologistic orientation, Marxism explains historical evolution by variation in the methods of production, in the forms of social organization, and in the organization of knowledge—variations in the relatively external "mediations" between human beings and nature. Variations in these factors are ultimately "selected" because of the superiority of more advanced productive forces in satisfying human needs.

## Alienation and racism

Such a view of the progress of human thought contradicts IQ theory because the latter abstracts from this development, both by its method of measuring knowledge and by the theory of knowledge which is incorporated into test construction. This theory identifies real intelligence with an individual's *unaided ability* to perform the most primitive types of formal exercises. Such a concept of human psychology is no mere error on the part of psychologists. It corresponds to the essential features of social organization under capitalism. The "plausibility" of IQ theory stems from the fact that the primary means of human development are privately owned, and seemingly external to the existence of the individual. The presupposition underlying IQ theory is not only theoretical but practical. It consists in the real, socially organized separation of the individual from the technical, social and intellectual conditions of existence. As a result of the private appropriation of the objective conditions of human existence—technical, social as well as intellectual—the real human essence, defined by social labor, takes the form of an external necessity, imposed upon individuals as a painful means for preserving mere biological existence. "As a result, therefore, man (the worker) only feels himself freely active in his animal functions—eating, drinking, procreating, or at most in his dwelling and in dressing-up, etc.; and in his human functions he no longer feels himself to be anything but an animal."[14]

The plausibility of IQ theory, which situates intelligence "within the skin" of the individual stems from the sociological experience described and explained by Marx in his analysis of the psychological consequences of practical "alientation." Fundamentally, alienation consists in the appropriation of the external conditions of human self-realization by the capitalist class, and the fact that the "reconnection" of the worker with these means of existence—necessary for the life of the species—takes place primarily for the purpose of extracting surplus labor for the capitalist. The worker tends therefore to see work, not as a means of self-realization or humanization, but as brutalizing. He tends to look for his own realization outside of the work process, in immediate quasi-biological forms of activity.

Marx vividly describes the precarious state of being of the individual whose relation to productive activity in the form of employment is constantly threatened. The connection with the means of earning a living does not appear as a human right or as part of the essential life activity of the individual member of society, but as an accident. "There is the production of human activity as *labor*—that is, as an activity quite alien to itself, to man and to nature, and therefore to consciousness and the flow of life—the *abstract* existence of man as a mere *workman* who may daily fall from his filled void into the absolute void—into his social, and therefore actual non-existence."[15]

The tendency for the individual to locate his essential being in immediate quasi-biological forms of existence has therefore both positive and negative features for the worker himself. On the one hand, it is outside of his work activity that he is "at home," while in his work he exists not for himself but for "another." On the other hand, this alien relation to the work activity constitutes a perpetual threat to the worker. He is threatened constantly with "falling from his filled void," into "social non-existence" and real starvation. This is the negative side of the sphere of life outside of the work place. This is the negative side of immediate, individual and animal-like existence.

This practical situation is at the root of the complex features of biological trends in ideology, including racism and male supremacy. Since all that the worker has to sell is basically his or her muscle and brainpower, it seems "natural" to attribute success in finding work to some innate capacity, unrelated to the logic of social and economic development. On the other hand, the social opprobrium and fear connected with "social non-existence," with the most unstable and brutalized forms of labor or with existence basically limited to the home, also appear to be due to innate biological features of the individuals concerned. Actually, such ideological tendencies connected with class society and especially with capitalism are realized in concrete, more specific forms depending on the particualr conditions obtaining in any country and during any specific period.

We have tried to outline the practical, experiential roots of biological and especially racist ideologies in the structural features of capitalism. Such tendencies are constantly reinforced by explicit racist ideologies and practices. These play on the sense that the individual worker may have that success or failure is the result of some quality, positive or negative, "inside one's skin." At the same time, countervailing tendencies are at work which lead to the recognition that in a state of isolated individuality one is powerless, and that only through the economic and political organization of workers will they gain real control over their destinies. In this case too, explicit ideological and scientific theory is needed to clarify the essential logic of human development and to expose the ideological and practical roots of racism.[16]

Marx criticized a psychology which describes the human individual as much as possible in terms of isolated, animal-like existence. Whereas the labor of humanity has produced an abundance of goods and talents, capable of enriching the lives of individuals, the private appropriation of the conditions of production and their realization for the profit of the few lead to an actual impoverishment of individuals, materially, socially and culturally, in the midst of plenty. Psychology takes this impoverished individual—stripped of the objective conditions of life but still basically producing them and being transformed by them—as the basis of its study. "We see how the history of *industry* and the established *objective* existence of industry are the *open book* of *man's essential powers,* the exposure to the senses of human *psychology.* Hitherto this was not conceived in its inseparable connection with man's *essential being,* but only in an external relation of utility . . . A psychology for which this, the part of history most contemporary and most accessible to sense, remains a closed book, cannot become a genuine, comprehensive and *real* science."[17]

The biological model in which the environment appears to be external to the individual and selects individuals according to variations rooted in organic differences in fact reflects the historical conditions of alienation under modern capitalism. Engels too writes of the estrangement of man's socio-historical essence from his immediate existence. This produces the appearance of a reversion to biological laws of development, which mask the essential social laws that continue to operate. As a result of human advances over nature, he writes,

> We have . . . infinitely multiplied production, so that a child now produces more than a hundred adults previously did. And what is the result? Increasing overwork and increasing misery of the masses, and every ten years a great collapse. Darwin did not know what a bitter satire he wrote on mankind, and especially on his countrymen, when he showed that free competition, the struggle for existence, which the economists celebrate as the highest historical achievement, is the normal state of the *animal* kingdom. Only conscious organization of social production, in which production and distribution are carried on in a planned way, can lift mankind above the rest of the animal world as regards the social aspect, in the same way that production in general has done this for mankind in the specifically biological aspect.[18]

In other words, the qualitatively new developments that have biologically distinguished man from the animal have yet to be fully realized in an historical form. Social production continues to evolve as a blind force—like a force of nature—and individuals continue to be subject to this process, passively adapting themselves to its eventualities, even when this apparently external environment is essentially the result of their own activity. But because this activity is privately appropriated, not fully comprehended and consciously coordinated, it seems to be a blind, externally determined or instinctive behavior. On the other hand, the tremendous developments of the economic, social and cultural wealth of modern society inevitably conflict with the form of social organization based on private appropriation and anarchy of production. Mankind struggles to reappropriate the conditions of its own development.

# 8

# Relative and absolute differences

*IQ* theory attempts to give plausibility to the *a priori* assumption
that intelligence is an essentially biological capacity unequally
distributed in the population. To do this it selects items which *seem*
to be unlearned while it tests individuals in such a way as to appear
to make them use their individual unaided intellect. Turning back to
the main features of IQ tests we can now recognize more clearly, not
only their *a priori* and unscientific character, but the way in which
they incorporate uncritical philosophical and social categories having their basic roots in the capitalist mode of production.

## Blind relativity

One of the main features of IQ tests is the fact that intelligence is
"measured" by variations from an average performance for each
chronological age. Such a form of measurement gives plausibility to
the concept that intelligence does not really develop. Thus while a
child progresses in knowledge and intellectual skills from one year
to the next, his or her "intelligence" is defined only in relation to the
average performance for the particular age group. Thus while one
ten year old may be "more intelligent" than another, it would be
incorrect, technically speaking, to say that ten year olds are more
intelligent than five year olds. It is of course useful and meaningful
to adopt a relative definition of intelligence. When we say that a
certain child is "bright" we mean "bright for his age." Teachers in
the classroom regularly deal with such relative judgments within an

age group. But most teachers probably think that they themselves are more "intelligent" than their pupils. IQ theory and methodology fixes one particular meaning of "intelligence" and excludes another. In this way, because of the relative stability of rank order in the classroom, a child's "intelligence," as defined by IQ tests, remains approximately the same even while the child is becoming more "intelligent" in the sense of quantitative and qualitative improvement of knowledge and skills.

This method of "measuring" intelligence is recognized to be quite peculiar, although the implications of the special character of statistical measurement are not clearly spelled out. Thus, Herrnstein writes that IQ is a measure of a person's relative standing in a group. Beyond that, it is not clear whether we know what we are actually measuring:

> A person's IQ is a different sort of fact about him than his height or his weight or his speed in the hundred yard dash, and not because of the difference between physical and mental attributes. Unlike inches, pounds, or seconds, the IQ is entirely a measure of relative standing in a given group. No such relativism is tolerated for the conventional measures. Gulliver may have looked like a giant in Lilliput and a mite in Brobdingnag, but he was just about seventy inches tall wherever he went. Relativism is tolerated for the IQ because, first of all, we have nothing better. If the testers came up with something like a platinum yardstick for mental capacity, it would quickly displace the IQ.[1]

This apparent modesty about IQ is soon belied by Herrnstein's attempt to turn this relative score into something absolute. We will show that Jensen attempts to perform the same "sleight-of-hand." But first let us be aware of the implications of this relativity of the IQ score. We understand that an IQ is a score which is related to an average performance *whatever that happens to be.* Two groups may differ enormously in real abilities, and yet the members for each group can be measured not by what they actually achieve but by how they stand in relation to the average score for their group. To measure Olympic swimmers and swimmers from the city of Buffalo in this way we would time the performances for the individuals in each group and then find what the average performances for the respective groups happen to be. Presumably the differences between the two groups would be considerable. However in our "relative" approach we would define the standard of measurement as the average for the particular group and assign that performance the score of 100. Individuals would then be measured by statistical techniques in relation to that average score of 100. An above average Olympic performer might score, say 130 (in the top 2%), while another would score 70 in the bottom 2%. The same calculations would be made for the "population" of Buffalo swimmers. As a result of this method of measurement the average swimmers for each group would have an *identical* "swimming quotient" (SQ). Could we assume from this that there is something "in" each swimmer which is the same, some "inner capacity" to swim which is measured by our swimming quotients? Could we say even that the inner *potential* of the average Buffalo swimmer is the same as that of the average Olympic swimmer, if only the right environmental conditions would allow for this capacity to be realized? Such an hypothesis would of course be sheer speculation based solely on the properties of a statistical procedure. One cannot deduce from the relative standing of an individual in a group anything about that individual's capabilities unless we are able to measure performance by some other criterion than relation to an average. In the present case we can measure racing speed and so are not likely to confuse the average Olympic swimmer with the average Buffalo swimmer.

The main point here is that such measurements are relative to a given group and hold only within that group. For example, while weight may not correlate with the performance of the Olympic swimmers, there may be a strong negative correlation among Buffalo swimmers, with overweight swimmers tending to do worse. The explanation of why weight correlates in one case and not in another has to do with the differences between the two groups. For obvious reasons few of the Olympic swimmers would be overweight and unfit—while the opposite may be true of the general population of Buffalo. However, it is important to recognize that it is statistically possible to measure distributions in each group in such a way that they take the same mathematical form. Although variation in poundage among Olympic swimmers may be much less than that of Buffalo swimmers, both groups would have an average weight, a top 10%, bottom 10%, etc. Only because we can actually weigh the individuals can we point out the objective differences between these two measures.

The fact that there is relatively little variation in the actual weights of the Olympic swimmers compared with the greater variation of the weights of the Buffalo swimmers is recognizable only when we have a measure of weight that is independent of the relative statistical measure. The lack of variation in actual weight for the Olympic swimmers leads to "restriction of range" which makes for less likelihood of correlation. Herrnstein gives a good example of this peculiarity of the correlational method when he points out that among a group of individuals with perfect pitch, there would be no correlation between differences in their musical ability and pitch, for the simple reason that their pitch is the same. One could not conclude that perfect pitch had nothing to do with musical ability, but only that it cannot explain the *differences* in ability or performances.[2] This conclusion should be applied to IQ tests as well— although Herrnstein himself does not make this point. Thus items are selected on the basis of their correlating with various other estimates of intellectual differences. Items or kinds of skills which all children perform equally well, and which do not vary, would not correlate with intellectual differences, and are excluded from IQ tests for the particular age group. This does not mean that they have no causal connection with intelligence or intellectual performance. The exclusion of such items or skills from the definition of intelligence, simply because they do not explain intellectual differences, is like excluding perfect pitch from a definition of musical ability because, in the case of individuals who all possess this ability, it does not explain their differences. Whereas we would not explain musical ability uniquely by *differences* in musical ability, IQ ideology, on the contrary, defines intelligence uniquely in terms of intellectual differences.

This brief discussion of statistics is important for a proper understanding of IQ. For with IQ we only know the distribution around the average, based on the statistical form of the normal curve. An average ten year old has the same IQ as an average five year old. Does this fact give us any more grounds for deducing some identical inner "something" in each child? The identity between the two children is not an identical inner capacity, but an identical relation to their respective groups.

In his *Logic,* Hegel discussed the way in which mathematical method can equalize qualitatively different objects. Writing of mathematical differences Hegel wrote that "quantitative difference is only the difference which is quite external. Thus, in geometry, a triangle and a quadrangle, figures qualitatively different, have this qualitative difference discounted by abstraction, and are equalized to one another in magnitude."[3] The same concept applies to the qualitative differences between five year olds and ten year olds. By applying the statistical method of measurement of differences by variation from the average in the normal curve distribution, one can "have this qualitative difference discounted by abstraction." As a result the differences "are equalized to one another in magnitude." Whereas the triangle and the quadrangle are equalized in terms of amount of space, the five and ten year olds are equalized in the character of the distribution of scores and in the stability of scores over time. The first "equalization" comes from the use of the bell curve as an *a priori* condition of test construction. The second, we have argued, comes from relative stability in the social and educational conditions of the child's development under normal circumstances.

Just as there is little danger that quantitative, statistical methods of analysis will lead us to confuse the Olympic and the Buffalo swimmers, so the equalization of the triangle and rectangle in terms of spatial extension is unlikely to lead us to confuse the two different geometric forms. The reason is that in addition to the statistical and quantitative method of "equalizing" the different phenomena we also have methods of measuring their specific features. However, as Herrnstein points out, we have no such "platinum yardstick of mental capacity." The raw scores on IQ tests, which are converted into IQ scores, have no "absolute" meaning comparable to measures of actual performance such as swimming speed. These raw scores themselves presuppose the relative measuring approach of IQ test construction. Items are chosen with a view to the differentiating effect they have on the population. An excess of items that produce the same effect would be redundant. A vocabulary raw score, for instance, has no necessary relation to amount of vocabulary an individual possesses. Words are selected for their differentiating effect at each age group.

Herrnstein regards Binet's originality as consisting in the fact that instead of finding a theoretical definition of intelligence for constructing tests, he took the "pragmatic" approach of finding norms for each age which correlated with evaluations of brightness or dullness made by teachers, etc. But is what is "bright" for a five year old the same as what is "bright" for a ten year old? Has this method grasped a common reality to these relative judgments, which would be the result of identical inner capacities? Herrnstein puts the question appropriately: "As Binet well knew, the chronological approach to intelligence finessed the weighty problem of defining intelligence itself. He had measured it without having said what it was. It took a while to know whether the sleight-of-hand had in fact yielded a real intelligence test or just an illusion of one."[4]

Contrary to Herrnstein, we believe that the sleight-of-hand is indeed an illusion produced by the combination of statistical abstraction and the fact of relative stability of rank order in school. No innate or fixed capacity can be legitimately derived from this. Thus an IQ score by itself says that an individual is above or below average without being able to say what abilities or performances follow from or are the basis of this fact. Were IQ merely a rank order of abilities and performances for a given age, such abilities could perhaps be described, though not necessarily explained. But for Jensen IQ is supposed to be a measure of basic mental capacity. There is no "platinum yardstick for mental capacity," however, because "mental capacity" is itself a fiction constructed on the basis of IQ tests. Thus only relative standing—or "relativism" is in fact left. From the materialist point of view, such scores which rank order individuals without being able to explain the content of that rank order, are essentially unscientific. Soviet psychologists, Teplov and Nebylitsyn criticize the "blindness" of Western "testology":

> A significant number of tests represent "samples" obtained in a purely empirical way, whose significance must be established statistically. We believe, however, that if the physiological or psychological meaning of a test is not clear, if a test is "blind," then no statistical processing will be able to yield scientifically reliable results. "Blind tests" cannot acquire a sense of "vision" merely by the application of statistical methods.[5]

### Basic historical differences

The discounting of qualitative differences by abstraction, coupled with no supplementary method for examining these qualitative differences, is revealed more clearly if we extend our view of intelligence beyond the comparison of age differences for a single individual to the comparison of differences between epochs. As Engels wrote a hundred years ago, " . . . a child now produces more than a hundred adults previously did."

The same differences could be revealed again today between what a ten year old can do now, and what a child of the same age could do when Engels wrote. The longer historical perspective places more obviously into question the biologistic perspective displayed by Herrnstein when he summarized Binet's discovery as the idea that "age confers intelligence." We have already examined the way IQ ideology attempts to create the illusion of measuring innate capacity by a predilection for choosing tests that involve general logical skills which have existed in a clearly formulated way for 2,500 years. The following example will bring out more sharply how the same "sleight-of-hand" is accomplished through exclusive reliance on the normal curve to measure intelligence.

Let us imagine an attempt to measure a hypothetical "calculating quotient" (CQ) which is meant to measure innate calculating ability. To achieve such a measure we would have to use items that approximate a common sense view of what basic calculating ability consists in. Suppose we decided that such raw calculating ability consisted in computing numbers in one's head. We could then devise a test of this ability which would differentiate individuals in the form of the normal curve distribution. We would have an average performance and variation above and below the average. This would be true of any group, although to make our study "representative" we would take a representative sample from the U.S. population as our base. The main point is that we are concerned here not with how well one can calculate, but with how one's calculating ability stands in relation to the average, whatever that happens to be. Rapidity of calculation, rather than accuracy, might be the simplest way to differentiate individuals in relatively simple problems of arithmetic.

An historical examination of our choice of test item or type of skill reveals, however, that "calculating in one's head" requires the use of a definite method of calculating, using historically evolved number and decimal systems. Other methods of calculating might produce different results. The use of the abacus is more primitive even though one is not just "using one's own head." But when we think we are "using our own head," we are in fact using a definite symbolic system which evolved externally to the evolution of one's head, and is just as much a "tool" of calculation as is the abacus. We chose "calculating in one's head" rather than calculating with pencil and paper, because the former seemed to approximate much more closely to a common sense notion of "pure" calculating ability. However, if calculating in one's head involves using a tool, why not avail ourselves of the normal tools of calculation, pencils and paper? In fact, calculating did not begin with calculating in one's head at all, but with calculating on one's fingers, hands, legs, head, etc. Primitive methods of calculating involved complex systems of using one's body for counting. A test could be devised for such a calculating ability in which the contestants would have to count on their fingers. The result would be the same normal curve distribution however. The following chart plots two methods of measuring calculating ability, one which uses the "relative" method of statistical measurement and the other which uses an absolute measurement of the real range of calculation permitted by each method.

I. Relative CQ

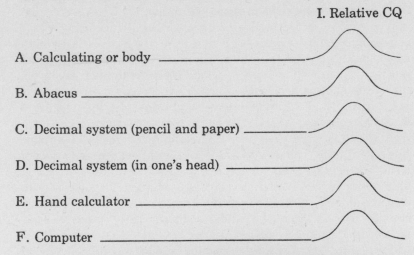

A. Calculating or body

B. Abacus

C. Decimal system (pencil and paper)

D. Decimal system (in one's head)

E. Hand calculator

F. Computer

The graph measuring "absolute calculating ability" plots a hypothetical average performance in real calculation based on the variations in methods of calculating that have evolved over time, and which have evolved with great rapidity during this century. The first method of measuring the qualitatively different kinds of cal-

## II. Absolute Calculating Ability

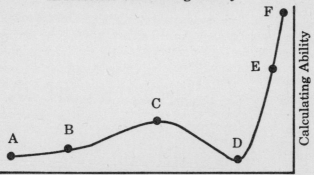

Historical Development to Infinity

culating ability shows no difference between them because the statistical method "discounts these differences by abstraction" and equalizes them. The second form at least measures the absolute progress in calculating abilities achieved by the different methods. The generally progressive development is only interrupted by the introduction of the method of calculating with the use of the decimal system of arabic numerals, but without the use of pencil and paper. Such a drop in ability is not due to the fact that raw, biological calculating ability is approximated, but is caused by our hypothetical mental tester who has deprived the testees of the physical tools of calculation which normally accompany the intellectual tools.

This example shows that the exclusive use of the normal curve statistical form "discounts in abstraction" the real historical differences in actual calculating ability. At each level, of course, there are variations in the mastery of that ability, and it is a legitimate question to ask what causes these differences. However the differences in real ability that result from the progressive evolution of new methods of calculation makes these differences at each level appear secondary if not completely trivial. Thus the ancient Egyptian high priest, who occupied the far end of the curve of differences of his period, would be totally overpowered in terms of real abilities by the much more average skills of the calculating merchant of subsequent times, (to say nothing of today's average ten year old equipped with a hand calculator).

No doubt a scientific study of intellectual development would reveal connections between the most primitive forms of calculating in individuals and eventual mastery of more abstract, more intellectual and scientific methods. IQ methodology does not, however, study the real mechanisms and conditions of intellectual development. Instead it employs a method which, in form and content, attempts to approximate a preconceived notion of the functioning of the unaided brain. In this it simply reflects unhistorical and undialectical concepts of thought as the activity of the isolated individual stripped of historically evolved methods of social production.

## Genius and history

But are all differences that might be due to biological variations essentially trivial ones? Are there not exceptional individuals, geniuses, whose contributions to mankind are noteworthy and whose excellence cannot be explained in any other way than by reference to their innate qualities? The biologistic theory of genius stands in sharp contrast to the historical materialist theory of development, which, while giving full credit to the contributions of great individuals, sees the basis of individual "geniuses" in the inner potentialities of their historical period—not in their biological substructure. Based on a biological theory of exceptional individuals, those individuals falling in the upper end of the normal curve, Jensen writes:

> It may well be true that the kind of ability we now call intelligence was needed in a certain percentage of the human population for our civilization to have arisen. But while a small minority—perhaps only one or two percent—of highly gifted individuals were needed to advance civilization, the vast majority were able to assimilate the consequences of these advances. It may take a Leibniz or a Newton to invent the calculus, but almost any college student can learn it and use it.[6]

Jensen assumes that intelligence created modern civilization, and several times refers to individual genius as an "argument" for a biological theory of intellectual differences. Jensen assumes a hierarchical "meritocracy" which begins with the gifts that come from on high, out of the innate intelligence of exceptional individuals, and filters down to a second level, which by no means includes everyone. If Newton and Leibniz invented the calculus, it takes "college students" to learn and use it, while the "vast majority" are able to "assimilate the consequences." The pyramid model implied here stands in contrast to the humanist concept of historical materialism which places the struggle of the masses of people at the basis of the developments which reach fruition in the minds of great individuals—who are great because they are able to draw from the wellsprings of the greatness of their times.

Engels describes some of the features of the revolutionary epoch that underlay the works of Newton and Leibniz:

> Modern natural science—the only one which can come into consideration *qua* science as against the brilliant intuitions of the Greeks and the sporadic unconnected investigations of the Arabs—begins with that mighty epoch when feudalism was smashed by the burghers. In the background of the struggle between the burghers of the towns and the feudal nobility this epoch showed the peasant in revolt, and behind the peasant the revolutionary beginnings of the modern proletariat, already red flag in hand and with communism on its lips. It was the epoch which brought into being the great monarchies in Europe, broke the spiritual dictatorship of the Pope, evoked the revival of Greek antiquity and with it the highest artistic development of the new age, broke through the boundaries of the old world, and for the first time really discovered the world.
>
> It was the greatest revolution that the world had so far experienced. Natural science also flourished in this revolution, was revolutionary through and through, advanced hand in hand with the awakening modern philosophy of the great Italians, and provided its martyrs for the stake and the prisons.... It was a time that called for giants and produced giants, giants in learning, intellect and character . . . .[7]

The historical investigation of the development of human thought cannot be separated from the basic development of mankind, a development that has its roots primarily in the struggles of great masses of people, classes and nations. The individual of genius grows up in a living historical context, and probes the needs and the possibilities of his times, on the basis of the practical and cultural achievements that have preceded him. To understand both the form and content of his thought it is necessary to know more than the abstract fact of the individual's exceptional achievements. The fact that Newton's achievements were in particular areas of science, and not in others, was not a result of some limitation in Newton's genes but is related to both objective and logical necessities in the development of science.

Even the general fact of exceptional individual genius has a limited historical character to it. As the contemporary revolution in science and technology develops, the character of intellectual activity changes as well. The fact that access to the "means of mental production" must be broadened both quantitatively and qualitatively points to an end to the historical division of intellectual and manual labor in which only a few individuals can devote themselves to intellectual specialization, especially of a creative kind. Moreover, intellectual developments become more and more the direct result of collective undertakings on an international scale. While breakthroughs and theoretical revolutions may still occur in one place rather than another, through the hypotheses and findings of one individual rather than another, these individual accomplishments are more and more obviously the consequence of the joint efforts of many individuals. And the "explanation" of the creative achievements of humanity by the mystique of biological genius shows itself to be not only empty of real scientific value for an understanding of the conditions of intellectual development, but part of the anti-social obscurantism that still impedes the full development of knowledge.

Thus a deeper understanding of the accomplishments of great individuals reveals the potentialities of human development at a time when, because of the alienation of labor, these potentialities become fully realized only in a few exceptional individuals. Marx wrote that "the exclusive concentration of artistic talent in some individuals, and correlatively its stifling in the great mass of people, is a consequence of the division of labor."[8] Much psychological theory attempts to analyze the individual in isolation from the real social relations in which one develops because the division of labor in class society, and especially under capitalism, really tends to isolate the individual from the full social conditions of development. Private ownership of the means of production (mental and physical) restricts the great majority of the people to participation in a fragment of the wealth of potentialities objectively created by society. While the individual who expresses something of this real historical potentiality in one domain or the other appears, on a empirical level, to be the exception to the rule, a deeper understanding of historical development shows that this individual is actually closer to the real norm of his time. It is the *suppression* of such development among the majority of the population which should be, and increasingly is, regarded as exceptional and abnormal.

## Population: concrete or abstract?

Despite the acknowledged relativity of IQ methods of measurement, IQ ideologists nevertheless attempt to deduce from them some absolute capacity. Herrnstein uses an "argument from the size of the population" to eliminate as much "relativity" as possible from the IQ tests.

To give some further appearance of plausibility to the idea that IQ tests are real measures of intelligence, test construction is based on a representative sampling from the population of an entire country. Moreover, tests standardized in one country can be used in some others without substantially losing their capacity to "sort people out" in a manner that correlates with "success" and "failure" in school and life. Herrnstein writes that the child is therefore compared, not with some small group but "effectively with the entire population of Western society. . . . And if it can be assumed that so large a sample of mankind is reasonably representative of the whole, then a relative measure is quite informative. An IQ of 100 would then indicate average intelligence, compared to people in general and not to some small group."[9]

The assumption that the larger the population the more meaningful is the relation to its average is equivalent to the quite doubtful statement that we know more about a person who is an above average Buffalo swimmer than one who is an above average Olympic swimmer. The larger and the more heterogeneous the population the more abstract becomes the IQ method of comparing the individual with the average. IQ tests in fact presuppose that the population which is measured can somehow be equalized or regarded as homogeneous, so that the differences in performance can plausibly be attributed to differences that are within the individual. In fact the opposite is the case. The more precisely we can define the group, the more meaningful are the averages of that group. Thus it is more meaningful to say that a child is an average seven year old than that he is an "average child." And it is more significant to say that the child is an average working class child of seven than simply an average seven year old. The more we know about the group the more content can be attributed to its averages, and the variations around those averages. Herrnstein's statement that although IQs are relative, they enable us to compare an individual with "people in general," is a further example of the way in which IQ methodology discounts qualitative differences and substitutes the most abstract formulations.

In fact, as Marx wrote in his discussion of method in political economy,[10] a "population" is only superficially a concrete object of study. On deeper analysis it turns out to be more than a collection of individuals who can be "equalized" by the same standard of measurement. A "population" is made up of definite social classes, whose relations, under conditions of private property, are antagonistic. The "population as a whole" is an abstraction from these definite socio-economic classes at a particular historical level of development. Herrnstein's assumption that comparison with the population of "Western society" is so meaningful as to be "reasonably" equivalent to comparison with "people in general" clearly betrays the abstract, ahistorical conception of human nature that underlies Herrnstein's attempt to make IQ tests seem valid objective measures of "something," despite their relativity.

### Jensen: from relativism to absolutism

Herrnstein's sleight-of-hand in turning the "relative" IQ score into something "significant" (and not "blind") is less obvious than some of the maneuvers performed by Jensen to achieve the same end. Thus Jensen acknowledges the relative character of IQ scores: "intelligence test scores are not points on an absolute scale of measurement like height and weight, but only indicate the individual's relative standing with reference to a normative population."[11] Jensen also acknowledges that the use of the normal curve is an *a priori* condition of test construction, and asks "But the important question still remains to be answered: is intelligence itself—and not just our measurements of it—really normally distributed?"[12] The *a priori*, operationalist definition of IQ around relative norms raises the question as to the objective validity of IQ as a measure of innate capacity. It is untrue, Jensen argues, "that intelligence exists only 'by definition' or is merely an insubstantial figment of psychological theory and test construction. Intelligence fully meets the usual scientific criteria for being regarded as an aspect of objective reality, just as much as do atoms, genes, and electromagnetic fields. Intelligence has indeed been singled out as especially important by the educational and occupational demands prevailing in all industrial societies, but it is nevertheless a biological reality and not just a figment of social convention."[13]

This passage reveals Jensen's awareness of the difficulties of deducing an absolute characteristic from a relative norm, as well as his adherence to a Social Darwinist conception of intelligence as a "biological reality" and not just a product of social "convention."

We have already examined one attempt to turn the relativity of IQ scores into an absolute measure by factor analysis and intercorrelations of factors. While Jensen believes that Spearman's "g" is the "rock of Gibraltar" in psychometrics, we have seen that the intercorrelation of test items is also an *a priori* condition of test items. Jensen performs a sleight-of-hand when after calling "g" a hypothetical construct and cautioning against "reifying" it (i.e., regarding it as real) goes on to call this "rock of Gibraltar" the source, albeit hypothetical, of intellectual differences, and the fluid core of crystallized intelligence.

Since Jensen's method consists of compiling a string of arguments "no one of which is convincing," we will examine some other attempts to turn the historically relative and blind IQ score into something absolute, ahistorical and, so, plausibly biological. Jensen solves the problem of the relativity of IQ scores quite simply in another context by *declaring* them to be absolute. Thus in his dicussion of the desirability of raising the intelligence of the population, and the dangers of "dysgenic trends," Jensen shows his awareness of the relativity of IQ tests, while offering a unique solution to the problem:

> If the intelligence of the whole population increased and our IQ tests were standardized anew, the mean IQ would again be made equal to 100, which, by definition, is the average for the population. Thus, in order to speak sensibly of raising intelligence we need an absolute frame of reference, and for simplicity's sake we will use the *present* distribution of IQ as our reference scale. Then

it will not be meaningless to speak of the average IQ of the population shifting to values other than 100.[14]

On the one hand, Jensen admits that by *definition* there can be no raising or lowering of "intelligence" as defined by IQ test method. Tests are so constructed that the average successful performance, whatever it is, will be the norm or base for measuring the variations, both above and below it, each of which must represent 50% of the population. However he also recognizes that in order to compare two different populations, at two different periods of time, it is necessary to have some "absolute" scale of reference. We have argued that it is indeed possible, by methods other than IQ tests, to measure historical intellectual development, at least approximately. The standard of such a measurement can only be, in the last analysis, the level of development of scientific theory, validated in the long run by experimental method and practical application. Jensen, however, looks for the absolute standard of measurement *within* IQ tests. We will simply take today's IQ tests and measure other people by them. This results in the technically absurd statement that there will be populations whose average intelligence will be lower or higher than 100. Since *by definition,* and by the "operation" of IQ test construction, this can never be the case, Jensen solves the problem of the relativity of IQ tests by *changing the operation.*

Thus Jensen writes that "In a bygone era, when the entire population's work consisted almost completely of gathering or producing food by primitive means, there was little need for a large number of persons with IQs much above 100."[15] Technically this statement means that there once was a time when few people were above average for their time! Jensen recognizes that IQ tests constructed for a different time would have a different content. He nevertheless believes that since IQ tests today measure absolute mental capacity, it is possible to measure primitive societies by them. But to do this one would have to abandon the IQ method of constructing tests relative to the average successful performance for that society. Whatever one may think about the possibilities of applying IQ tests—presumably "performance" tests requiring no literate skills—to a primitive hunter-gatherer culture, Jensen has implicitly recognized that the IQ tests cannot be used to measure absolute intelligence without changing the nature of these tests.

We can deal briefly with some other arguments which Jensen

believes are not wholly convincing, but when added together, make the "genetic hypothesis" plausible. Thus Jensen argues that IQ scores "behave" analogously to measurements of physical properties for which genetic explanation of differences are not in dispute. For example, IQ scores begin to be predictable around age four and become fairly stable at about age eight. The same thing is true, Jensen informs his reader, with respect to physical height. That is, before the age of four one cannot predict with much accuracy that a relatively tall infant will become a tall adult. But by about the age of eight the rank order of children's height has become relatively well established. Jensen draws no explicit conclusion from this analogy, but the implication is there for those who are not biased by "dogmatic environmentalism."

Two different causes can of course produce these statistically analogous results. However, a better reply is evident once we step outside the IQ framework and look at intelligence historically. Real human intelligence has been advancing at an almost exponential rate, whereas increase in average human height has occurred on a much more modest scale. By viewing intelligence materialistically, as a real reflection of the world, and dialectically, as a reflection that becomes deeper and richer with time, we necessarily come up with a concept of intellectual growth that requires qualitatively different causes from those which determine physical growth. So much is this contrast evident, once we step out of the charmed circle of the IQ theory, that it does not seem an exaggeration to agree with Engels that, especially in the modern period, "the development of the sciences proceeded with giant strides, and, it might be said, gained in force in proportion to the square of the distance (in time) from its point of departure. It was as if the world were to be shown that henceforth, for the highest product of organic matter, the human mind, the law of motion holds good that is the reverse of that for inorganic matter."[16] Only by restricting our view of intelligence to the relativistic framework of IQ, can one compare what Jensen calls "development rates" of intelligence and height. When we do this we find that it is possible to measure both phenomena relatively, in relation to the prevailing average, and that rank order becomes relatively stable at about the same age.

Another comparison with height is made when Jensen argues that height is "normally" distributed, like intelligence, and so both "may" be subject to genetic mechanisms. Behind this argument is the notion that a trait such as height which is subject to the influence of many genetic factors will vary in the same population in a relatively "normal" way, with most people clustering around the average and with no sharp gradations, as is the case with traits, such as eye color, which are subject to the influence of a small number of genes. This argument cannot be applied to intelligence, however, because while we can measure height "absolutely" and *discover* a normal curve pattern in the distribution of heights, in the case of intelligence the process is just the opposite. Since no objective measure of "intelligence" is said to be possible, its normal distribution is *presupposed* to exist and made to exist as a result of test construction. We would assume that a real measure of intelligence in terms of actual knowledge and skills, including scientific theory at a relatively basic level, would discover quite a different pattern than that of the normal curve. We would expect, in accord with the continued existence of a division of mental and manual labor, to find sharp differences in the population in terms of theoretical knowledge, despite the fact that the brain is subject to far more complex genetic influences than height. Thus if we are measuring "algebraic capacity" or "knowledge of basic electronics" among sixteen year olds, for example, we would expect to find sharp differences in performance. Such differences relate to the unequal distribution of and access to the external social and intellectual "means of mental production."

Further examination of Jensen's arguments show continued sleight-of-hand to create the illusion that the relativist and *a priori* IQ definition of intelligence can produce an absolute, objective measure of intelligence. Thus Jensen attempts to overcome the difficulty involved in the recognized *a priori* character of the normal curve as a built-in condition of the construction of IQ tests. Jensen argues that if IQ scales behave as "interval scales" they can be said to measure something objective. As an example of an interval scale Jensen chooses the centigrade thermometer. We can predict that a mixture of equal amounts of water at 0° and at 100° will result in water of 50°. If IQ scores can behave in the same way, then we can be reassured that they are not fictions of IQ test construction, but reflections of real properties.

To produce such a mixture of IQs Jensen turns to studies of

inheritance of intelligence—a matter which we will treat in greater length shortly. Jensen argues that the "most compelling evidence" for the "objectivity" of the normal curve "comes from studies of the inheritance of intelligence, in which we examine the pattern of inter-correlations among relatives of varying degrees of kinship."[17] This turn to the "heritability" of intelligence shows how weak are the attempts to defend IQ tests as objective measures of intelligence by staying within the realm of IQ tests themselves. Jensen follows the method of "construct validation" by bringing in correlations with other data, presumably independent of IQ tests and of a totally different nature, to prove that IQ tests measure something "real," that is, biological.

Again Jensen argues by analogy with height, which follows the statistical pattern of "regression to the mean." What this means is that children tend to be "more average" than their parents in height. If we know how much above or below the average the parents are, it is possible to predict that the children will normally be only half as much above or below the average as their parents. This .50 correlation of the height of children with that of their parents is matched by an average .50 correlation of the IQs of children with those of their parents. Thus if the parents' average IQ is 115, their children will tend to have an IQ half way between that of their parents and that of the population average of 100, i.e., the children will tend to have IQs of 107. On the other hand, parents with an average IQ of 85 will tend to have children with IQs of 92 or so. Eysenck cites this "regression to the mean" phenomenon in a secondary dispute with Herrnstein. Although Eysenck agrees with Herrnstein that increasing environmental equality leads to a greater role for biology in assigning individuals their slot in life, he argues that regression to the mean prevents the formation of a rigid caste system. Instead of a tendency towards a genetic meritocracy, Eysenck proposes a tendency toward genetic "mediocracy."[18]

Does not this example prove that IQ measures something "real"? This, of course, we have not denied—only denying that this reality is "mental capacity." A response to Jensen's argument will once again show the importance of distinguishing between an "absolute," a historical conception of intelligence and a relativist conception of intelligence as meaning rank order in school success.

In the first place, it is important to note that there is something strange about the argument regarding height. It is well known that average height has tended to increase, and that children in general tend to be taller than their parents. How can one accept or explain the "statistical fact" that children of people of above average height tend to be smaller than their parents? But this line of thought has confused "absolute" and "relative" perspectives. We assume that the meaning is as follows. While parents may be exceptionally tall *for their generation,* their children may be *both* taller than their parents *and* "more average" in height *for their own generation.* "Regression to the mean" does not refer to absolute height but to relative height. Thus while the second generation may be taller than the first, "absolutely speaking," the statistical method of measuring height in relation to the average yields the same bell-shaped curve for both generations since their absolute differences have been "discounted by abstraction" and the two populations have been "equalized in magnitude" through the use of the same statistical form of measurement. The confusion between the two perspectives would lead one to think that because the statistical form of rank ordering individuals by height is the same for both generations there has been no growth in average height from one generation to the next. This confusion is easy to avoid because we can measure the actual height of the individuals in each group. But no such "platinum yardstick for mental capacity" is recognized for IQ measurement.

Thus, regression to the mean in IQ does not mean that children are less "intelligent," absolutely speaking, than their high IQ parents, but only that they are less exceptional for their own age group. In absolute terms, the children may have twice the theoretical skills and knowledge of their parents, but in relative terms they may be less exceptional for their own age group than their parents were for theirs. IQ measurement, which deals solely with relative standing, thus abstracts from the real, substantial developments of knowledge and mental skills that took place over the generation. Recognition of the difference between real "absolute" growth of intelligence, and relative standing for a given group should be comforting to Eysenck who dedicates his book *The Inequality of Man* to his children "in the hope that genetic regression to the mean has not dealt too harshly with them!"[19]

The fact that children of exceptional parents should be less exceptional for their own generation can be explained by arguments which do not have to bring in genetics. Regression to the mean expresses laws of probability which could apply to genetic as well as to environmental conditions. Thus, if we assumed that exceptional conditions lead to the development of exceptional parents, it is less likely that their children will have the same exceptional conditions. Instead, the children would tend to be a sort of cross between the fact of having intellectually exceptional parents and the fact of having to grow up nevertheless in the prevailing "averaging" conditions of their times.

## Difference and identity

The very measurement of "pure" differences makes assumptions which fail to stand up to dialectical and historical analysis. In his dialectical *Logic,* Hegel devotes considerable attention to the concept of difference and its logical counterpart "identity." Hegel criticizes the method of "abstract understanding" (which Engels calls the "metaphysical method"), which consists in attempting to isolate these concepts, to treat identity or sameness without difference, and vice versa.[20] IQ method involves an attempt to measure "pure" differences, both in the selection of items and in the statistical form of analysis. In fact, however, the tests presuppose a common culture, an "identity" or a relationship between the subjects measured. This common element is "discounted by abstraction" although it is everywhere presupposed, haunting the "differences" despite all attempts to eliminate it. The argument that by making environment "equal" we can separate the differences, and attribute them to genes, overlooks the fact that differences occur only in relation to a common element. Far from being pure differences, these are differences springing from a definite social-historical context.

The separation of "difference" from "identity" is reflected in the method of constructing IQ tests. As we have argued, items which all can answer are not "good" test items, and are excluded from a test whose main purpose is to differentiate the children. This is a methodological "operation" by which tests are constructed. Should we try to base a scientific definition of intelligence on such an operation? Or should we not recognize how one-sided is the concept of intelligence which is based on attempting to isolate differences? The insufficiency of the IQ definition of intelligence is evident in this attempt to exclude what the children have in common from the definition of their "intelligence."

It is a law of dialectical logic that polar terms such as "identity" and "difference" cannot be separated. The attempt to do so does not really succeed in eliminating the polarity which inevitably asserts itself "thoughtlessly" in the process of thinking. Despite the attempt to measure pure differences between children, a test is only meaningful on the basis of a certain *relationship* between the children, a *common* element which makes the differences possible. IQ method implicity reflects "the common element" both in the content and the formal requirements of the test. Thus tests are valid only on the basis of a certain range of common content for each age group, and is, strictly speaking, applicable only to the population on which the test has been constructed. However, the intent of the test is to reflect only the differences. Hence there is no real study of the content of the tests—i.e., the definite kinds of skills for which there are variations—nor of the definite structure and laws of development of the society and its educational institutions in which the differences occur. Despite the fact this common element "vanishes" in the final result—the distribution of differences above and below the average—it is nevertheless continually present as a "mute" or "blind" force.

The attempt to isolate "pure difference" is not possible without reference to the logically opposite concept, "pure identity." In the theorizing of IQ ideologists, there is continual reference to "common culture" to "equality of opportunity," to the identity of the environmental experiences which make possible the separation of the opposite "factor"—i.e., the supposedly non-environmental differences. The abstract character of the concept of differences is adequately matched by the abstract character of the concept of identity or sameness of the "culture" or "environment" which is supposed to allow for the separation of differences.

Another example of the way in which there is a "blind" oscillation between supposedly opposite and separate notions is found in the confusion between IQ as rank order and IQ as an "inner capacity." Technically, as we have seen, IQ is a relative concept. An individual is compared to others, and the IQ measurement is a comparative one. Difference only makes sense in relation to that from which something differs. The base from which differences are measured is the average performance. The meaning of the difference is therefore found in the relation to the average. But what is the content of the average? Technically, the average is the zero point from which scores are measured as "deviations" above or below. And we do not know what the real content of the average is. Thus, for it to acquire its meaning, the IQ score refers the individual to something which is essentially unknown. IQ is a relation which relates to an unknown. Under these circumstances, it seems "natural" to turn away from this pure relativism and to make an absolute out of the relative score. Thus one attempts to find meaning in the score itself despite the fact that this is technically impossible. The common sense meaning of IQ accords with this tendency to absolutize the relative meaning of IQ. Sophisticated understanding of the relativity of scores by professional psychologists leads to some debunking of "vulgar" interpretations—but frequently with the precaution that IQ tests are still somehow meaningful or useful. In Jensen, however, there is a clear oscillation and sleight-of-hand in which the "absolute" meaning of IQ (which accords with the common sense or "vulgar" interpretation) clearly wins out over the relativist meaning.

It should be noted that the way IQ is mathematically scored makes it easier to interpret the score as an absolute. Thus the use of 100 as a base, rather than zero, makes the average score appear to be "something" rather than "nothing." IQ scores are technically equivalent to percentile scores. But if one were to say that "an individual is in the 50th percentile," rather than that "the individual *has* an IQ of 100," it would be much more obvious that what is being measured is primarily standing in a group and not something *in* the individual. Thus, because of the attempt to isolate opposites "metaphysically," the relation between opposites, according to Hegel, remains "unthought" or "blind." Consequently, there is oscillation from one pole to another without the theoretical capacity to control this oscillation. This "capacity" is given, not by innate mental powers, but by mastery of the laws of dialectical thought. Engels wrote that the results in which science summarizes its experiences are concepts, but " ... the art of working with concepts is not inborn and also is not given with ordinary everyday consciousness, but requires real thought, and . . . this thought similarly has a long empirical history . . . " which must be successfully assimilated to be truly mastered.[21]

In fact, it is only the use of the same mathematical tool that enables mental testers to regard the differences that occur between qualitatively different kinds of skills as having the same form. In fact, the nature and significance of differences are themselves qualitatively different for each skill or level of cognitive development or for different societies or epochs. The differences between finger counting abilities do not have the same meaning nor necessarily the same form of distribution, as the differences between ability to calculate in one's head. We know that differences of height in a population are not the same kind of differences, either in form of distribution or practical consequence, as the differences between eye color. Because we understand something about the nature and causes of the trait in question we can explain why the differences themselves are different in each case. However, in the case of IQ tests, the form of the differences is the same at each age level and for each population only because of the *a priori* use of the same mathematical instrument of measurement as a condition of test construction.

The dialectical theory that opposite terms "interpenetrate" makes possible more concrete study of the interpenetration of identity and difference in each society and for different age groups. The general conception of a society as a definite kind of "unity of opposites" leads to the study of the interpenetration of unity and difference in each society. A class society is a unity of opposite and antagonistic classes. Any concept of "common culture" or "common environment" is the result of metaphysical abstraction and speculative argumentation. Socialist society, too, involves unity-in-opposition, but of a different kind. The nature and causes of differences here changes. The development of advanced socialism does not lead to abstract equalizing of environment, although it does lead to certain kinds of equalization—based in the first place on an equal relation to the means of production. The trend toward overcoming the inequality between mental and manual labor does not lead to a kind of abstract mental equality, however. On the contrary, greater social unity makes possible a richer, more "many-sided" development of the personality, skills and knowledge of individuals.

Differentiation is a *process,* not an abstract static fact. In dialectical theory early primitive forms of existence involve relatively undifferentiated forms of existence, while definite forms of existence develop through differentiation. This is a fact of the development both of the species and of the individuals in the species. Primitive human societies too have a relatively undifferentiated character in contrast to the later differentiation which occurs through social and technical division of labor and class formations. The differences that are of primary importance in primitive societies are therefore differences in such "natural" characteristics as age, sex and family relations. These differences are organized on the basis of common ownership of the means of production and of existence—primarily the natural means of subsistence which are appropriated through the activity of the collective. Individual differences have quite a special meaning and character in a society in which, as Marx wrote, the individuals have not yet cut the umbilical cord binding them to the tribe.

With the development of private property and the differentiation of society in terms of both greater division of labor and opposite class relations to the means of production, the nature of social differences becomes more pronounced while the real connections between individuals becomes concealed. The unity of class society is enforced by the state which, in a society divided by wealth, seems to constitute the sole basis of unity. Notions of "equality before the law" or "political equality" express the highest development of the state in class society—the bourgeois democratic state. Here, despite the fact that this ideal is never really achieved, individuals have their common *identity* through the political realm while *differences* are allowed to flourish in the "private" realm of the economy. Social Darwinist theory transposes the terms of this relationship to make it appear to be based on nature. For the Social Darwinist the "environment" is basically equal, while the private differences are basically biological. The economic realm of private differences—basically class differences—is transposed into the unalterable realm of biology. Thus, the focus on "differences," if theoretically unjustified, has its basis in the social structure of a society based on a certain social form of differentiation of individuals, where the differences seem to be more truly "real" than the sameness or identity which seems to be relatively external to the individual. In a society divided by class interests, based on the private production of commodities and wage labor, "common humanity" becomes an ideology of identity between exploiter and exploited. Marx explained the basis of this "cult of abstract man" in capitalist commodity relations.[22] This abstract unity can only appear ethereal—"heavenly" as Marx once wrote—in relation to the "earthly" differences that divide classes and individuals.[23] The difference between the "private" realm of the economy and the "public" realm of the state or politics results in the illusion that one can separate what is common or equal from what is different as though they were two different realms.

## Equality and inequality

Marx further concretizes the notions of identity and difference in his analyses of economic relations.[24] Here the categories of identity and difference, as well as statistical averaging tendencies, are understood through real historical analysis rather than through speculation with the use of abstract and unhistorical categories. We can only touch on some aspects of this analysis to show the difference between the use of these concepts in Marx's work and in psychometric theory.

Marx analyzes the exchange of commodities as involving a dialectical unity of equality or identity and difference. An exchange of goods presupposes an equalization of the goods, measured by a single quantitative standard. So many shoes, for example, are equal to so much fish. At the same time there is no point to the exchange unless there is a difference between the commodities. Qualitatively different goods are measured by the same standard representing quantities that are proportional to each other. In general, the equation between goods is represented in the form of money. Marx noted that Aristotle had long ago recognized the problem of what constituted the basis of this equalizing of different objects. Aristotle concluded that the relations of exchange, represented by money, were purely conventional.

Marx developed the thesis of the classical economists, especially Smith and Ricardo, that the standard of measurement of goods for purposes of exchange was not a social convention but the amount of labor needed to produce those goods. Marx stressed that labor, as a basis of exchange value, was the socially necessary labor at the *average* prevailing level of production for the time. Continual changes in the methods and manner of production, stimulated and imposed by competition, lead to fluctuations above and below this average. Thus, the development of a new method of producing cotton enables the individual capitalist to produce more cheaply relative to his competitors and gain an advantage in the competitive struggle of the market. A kind of Darwinian "survival of the fittest" occurs, in which the chief factor of survival is the cheapness of production, due to the productivity of the labor. This in turn, is the result of the methods of production, the intensity and quality of the work, and the wages paid to workers. Engels noted this external analogy between biological and capitalist laws of evolution:

> Darwin did not know what a bitter satire he wrote on mankind, and especially on his countrymen, when he showed that free competition, the struggle for existence, which the economists celebrate as the highest historical achievement, is the normal state of the *animal kingdom*.[25]

The means of "survival" in the capitalist struggle is not, however, the evolution of biological organs, but the evolution of more advanced "productive organs," outside of and separated from the direct producers.

The averaging process here is dynamic. It is stimulated by competition within the common competitive arena. Its ultimate content is the average prevailing level of production. In relatively equal competitive conditions, there is fluctuation around this average leading in an ascending direction, toward higher levels of average production. When an individual capitalist falls too far below the average rate of profit, he can no longer "survive" and tends to fall into the intermediate and subordinate classes.

For the capitalist, the direct measure of economic health, however, is not the level of development of the productive forces, but the more tangible and directly measurable rate of profit. The direct measure of economic development is the average rate of profit— around which individual rates of profit fluctuate. Because of the continual *growth* of the productive forces, however, the rate of profit, which is the ratio of amount invested in machinery, raw materials and labor to profits, tends to fall. This is a result of the fact that the proportion of "living labor" to "dead labor" declines with the very rise of productivity: A certain number of workers continually put into motion more and more machinery and raw materials, and the capital invested in them. Since it is the living labor that produces surplus value and hence profits, the ratio of profit to total invested capital tends to fall. This falling rate of profit is counteracted in many ways. State investment of capital is one significant method of counteracting the tendency for the rate of profit to fall. State investment permits even greater concentrations of productive forces and capital, and ultimately accentuates the problem of the rate of profit. A fundamental method of maintaining and increasing the rate of profit is through the export of capital to areas where it brings higher returns, chiefly because of the cheaper character of the labor, expressing inequalities in the developments of nations and the old colonial commercial relations. Imperalism does not simply find "underdevelopment," however; it produces and reinforces it. It does this through the imposition of force, the inculcation of a sense of inferiority in the subordinated populations, and the exacerbation of national chauvinist and racist ideology in the population of the imperialist nation.

The "competition between equals" takes place on the basis of growing inequalities between those fewer and fewer owners of the major means of production and the increasing numbers who have been proletarianized—forced to sell their capacity to work in order to earn a living. The growing power of the productive forces involves continual socialization of the productive forces as larger and larger numbers of workers must be coordinated within giant enterprises. At a certain point "competition between equals" itself leads to the differentiation of a few giant monopolists and the middle and smaller capitalists who fall significantly below the rate of profit sustained by the monopoly interests.

Not only is there a dynamic character to the "averaging process" stimulated by competition between interdependent units of production, but qualitative differences emerge between competitors. No abstract normal curve is able to reflect the dynamics of this process, although the concept of averages has real meaning and statistical methods have objective bases in real economic processes. However, to use statistical methods correctly they must be based on scientifically grasped categories and laws of development. Thus calculation of an average rate of profit for all private enterprises mystifies real economic relations. The fact that monopolies are able to maintain a permanently higher than average rate of profit substantially transforms the relations between property owners. Relations of domination and subordination here replace relations of equal competition—although such competition continues, in new forms, between the monopoly giants.

Marx exposes the fallacy of assuming economic equality between all individuals as owners of commodities. The capitalist and wage worker enter into a "contract" in which there is an equal exchange of "labor power" or capacity to work for a wage which tends "on the average" to be sufficient to allow that labor power to be reproduced. This equality or equalization is based on a unity of opposites—on a fundamental inequality between the capitalist who owns the means of production and the wage laborers who have nothing to sell but their ability to work. Thus, again, identity is possible only through difference, equality is dialectically related to inequality. This inequality or difference is qualitative, not just quantitative.

Between workers themselves quantitative differences fluctuate around an average wage reflecting the prevailing average level of development of the capacity to labor. This subjective capacity is based on the objective character of the means of production, which tends to grow to higher levels. Marx argued that the average prevailing level of skill could be regarded as a basis for the determination of wage levels. Marx called this "simple labor" although he did not regard it as an absolute, but as an average prevailing level of skill which continually changes with the development of the productive forces. While the real character of workers' abilities tends to grow, at any time an average prevailing level of culture defines the nature of "simple labor." "Complex labor" constitutes that higher level of skill which involves a degree of training significantly above the prevailing cultural norm for the majority of workers. On the other hand, the need for a reserve army of unemplcyed leads to a tendency to produce underdeveloped workers as well as populations. Racism, a product of capitalist society in general, is intensified by imperialism which, for the sake of super-exploitation, attempts to debase whole national groups. To justify this in a society whose property relations and ideology tend to commit it to some sort of belief in "the equality of man," the predatory expansion of capitalism fosters the ideology of the subhuman character of its most oppressed victims.

Thus within the working class and general working population there are variations above and below the average level of skills and culture. The division of mental and manual labor further complicates the picture of the development of skills. Marx described the historical process whereby intellectuals are torn between allegiance to the ruling class and alliance with the growing working class movement. Modern industry, with its requirements of large masses of intellectual workers, leads to closer relations between manual workers and the majority of intellectuals whose historically developed capacities are frustrated by conditions of work and the orientation of capitalist production based on pursuit of profit.

None of what has already been written can be applied to the basic difference in "capacity" between owners of the main "capacities" of social production and those whose subjective capacities to work with hand and brain are separated from these necessary external means of realizing as well as producing their "inner" capacities. While the concept of an averaging tendency is meaningful only where there is relative equality and competitive interaction between individual units, the concept of averages does not apply to the interaction of qualitatively different and opposed forces. To represent this opposition, Engels referred to the model of the "resultant" of opposite vectors.[26] Although he uses the concept of the resultant to represent the outcome of numerous interacting, but opposed forces—the resultant as an "average direction"—he basically refers to oppositions between classes which have opposed interests.

Whether in the use of the model of average tendencies, or in other mathematical models for reflecting real tendencies, the main focus for dialectical materialist methodology is a knowledge of the basic relations and laws of development of the field in question. Mathematical tools cannot substitute for basic conceptual analysis, but have to be subordinated to such real knowledge. IQ theory, without any inkling of the real laws of social and intellectual development, bases its methods on "common sense," i.e., ideological concepts of intelligence. The basic concepts of intelligence as an inner capacity, fixed at birth, selected by an external environment, distributed unequally in the population, and leading to social and economic differences, involve superficial "appearances" of society based on private property in the main means of production. On this basis, statistical methods, however refined, can only give a spurious appearance of scientific respectability to blatantly biased ideological imperatives.

# 9

# *Heritability*

*T*he history of IQ testing reveals the attempt to create an instrument that would measure innate and fixed intelligence. We have devoted the major part of this essay to an examination of the presuppositions of IQ ideology in order to explain how the *semblance* of fixed and innate intelligence has been derived primarily from the method of selecting test items and of interpreting scores. This effort is necessary for an understanding of just what IQ tests measure. In one respect, however, this effort may seem unnecessary. Jensen differs from his predecessors, Galton, Terman and others in the history of the biological theory of intelligence, in that he does not appear to *assume* the validity of IQ tests as a measure of innate intelligence. On the contrary, he pretends to *prove* this by use of the methods of "population genetics" which were unavailable in the early days of IQ construction. Jensen pretends to have effected a theoretical revolution in giving a genetic interpretation of intelligence and in challenging the reigning "environmentalist" interpretations.

## From innate IQ to IQ heritability

In fact the biological interpretation of intelligence not only was assumed from the beginning of the effort to measure intellectual differences, but has permeated the method of IQ construction. Nevertheless this biological interpretation has been challenged from various scientific directions. The rejection of eugenic theories and the biological interpretation of IQ scores grew with the revulsion against fascist thought in the 1930s and with the war against Nazi aggression in the 1940s. The newly formed United Nations sponsored international research to expose the fallacious and fundamentally vicious character of biological and racist theories of human society. It is a tribute to this collective effort of natural and social scientists, as well as to the general public rejection of Nazi racist theories and the untold destruction of lives which such theories justified, that the biological interpretation of IQ tests can no longer be assumed.[1] This is not to say that this belief does not still have roots in popular opinion as well as in theory, as we have noted in the beginning of this essay. There is still a widespread belief in IQ as a measure of one's basic capacity to think. But contrary beliefs prevent this tendency from resulting in the kind of conscious racist and nationalist attitudes that were deliberately cultivated in the twenties and thirties, and which were effectively challenged by the civil rights movement in the U.S. South during the late fifties and throughout the sixties.

Jensen's article in 1969 is aimed at reviving a biological interpretation of intelligence and of giving academic and scientific credence to the racist theory of the intellectual inferiority of Black Americans. For this revival, however, it was not sufficient for Jensen to report that whites score higher than Blacks on IQ tests. The mystique of IQ as a valid measure of innate capacity has been sufficiently challenged so as to make it necessary to prove validity on grounds seemingly outside of the IQ test itself. Thus, in addition to reporting racial differences in IQ tests, Jensen adds that by using methods of "population genetics" it can be shown that IQ is 80% "heritable." From there, Jensen marshals a number of arguments which render "plausible" the "hypothesis" that the main explanation of the differences between IQ scores for Blacks and whites is to be sought in genetic differences.

The appeal to heritability analysis of IQ scores is one further step in the attempt to validate IQ as a measure of innate and fixed intelligence. It is another attempt to find some criterion outside of the operational definition of IQ which would demonstrate that it measures "what it is supposed to measure." And yet, as with other attempts of this sort, it rests on *a priori* presuppositions that need to be critically evaluated. The presuppositions of heritability analysis are in fact very similar to those that we have examined in our study of IQ.

One obvious presupposition of the "heritability of IQ" is that it must begin with IQ scores. If IQ scores cannot be relied upon as measures of "general intelligence" then it follows that the "heritability" of IQ cannot be something meaningful. We think that our previous analysis has shown the essentially artificial and ideological character of IQ theory—especially as metaphysically and idealistically interpreted by Jensen. Nevertheless, if some further evidence is presented to validate IQ as a measure of innate intelligence, we should have to reconsider our earlier interpretations. To what extent, then, has "population genetics" come to the rescue of the much beleagured IQ theory?

## Heritability and heredity

The assertion that IQ is "80% heritable" gives the impression that a large part of our intelligence is inherited and owes relatively little to the "environmental" circumstances in which we have grown up, and there is not too much that can be done about changing the intellectual abilities that were conferred upon us at conception. In other words, the assertion is taken to mean that "intelligence" is *mostly,* but not absolutely, innate and fixed for life. Thus the expression appears to reassert, with somewhat more modesty, the main theses of IQ ideology.

On closer inspection, however, the concept of "heritability" has a particular meaning in population genetics which fails to support this "common sense" interpretation. The "heritability" of a trait does not in fact mean how much of some trait (the "phenotype") is due to the genes (the "genotype"), as opposed to the environment. Nor does it refer to something that is necessarily fixed; the "heritability" of a trait can change drastically. In fact the common sense interpretation and the technical usage operate on completely different planes. As we will see, Jensen is perfectly aware of the technical usage of the concept of heritability and of the incorrectness of the above "common sense" interpretation. Jensen displays sufficient expertise to convince biologists that he knows what heritability in fact means—while at the same time the main thrust of his argument is to reinforce the "common sense" impression that it is the measure of "how much of intelligence is determined by the genes." There is a sleight-of-hand, essentially similar to the one played in IQ theory, which confuses a technical expression, which has a peculiar *relative* meaning, with an intuitive belief that attributes an absolute, *metaphysical* meaning to the concept. The following examples will help to illustrate the use of the notion of heritability in biological theory.

Consider a situation in which there are two completely inbred lines of corn, line A and line B. In each line all of the seeds are genetically identical with one another (have the same genotype), but there is a genotype difference *between* the two lines. The seeds are planted in ordinary potting soil which gives a variety of different conditions for development. Since we know that in each line there is genetic identity, we know that the differences or variations in the height (the "phenotype"), say, of the corn in line A is due to differences in the *environment* of line A. Observed variations in the height of line B are entirely environmental in origin. Because none of the observed phenotypic differences (i.e., differences in height) in each line is due to differences in the genotype of the line, the "heritability" of each line is said to be 0%. Heritability is defined as the proportion of the total variation of a certain trait in a population which is due to genetic variation. It is the ratio of the genotypic "variance" over the total phenotypic "variance." In this example we know that there are no differences between the genes within each line. The technical formulation for variation is called the "variance" and is derived by a method which will remind us of IQ. We will return to this shortly. Since there is no genetic variation or differences, our formula for the heritability (H) of each line will be 0/t ("t" being the total variance) or 0%. The total variance is defined as the *sum* of the genetic variance and the environmental variance (t = g+e). We will return to this general formula for closer scrutiny.

While the heritability of both lines A and B is 0%, lines A and B are nevertheless genetically different from each other. Suppose that we observe a difference between the average height of line A and that of line B. We would probably be safe to assume that the cause of this difference between lines is due to the genetic difference between A and B. This leads to the paradoxical conclusion that while the "heritability" of each line is 0, the explanation of the differences between the lines is entirely genetic. Thus, there is no logical connection between the heritability figure for a given "population" and the cause of the real differences between two populations. The heritability estimate is only valid for a given population. In this it is also similar to IQ, which is entirely relative to the population on which it is standardized. This point is important on technical grounds because Jensen, while aware of the technical limitations of heritability, attempts to explain the difference *between* Black and white populations from the heritability figure that is said to obtain *within* each population. T. Dobzhensky points out the contradiction in Jensen's work between his awareness of the technical limitations of heritability and the conclusions he attempts to draw which overstep and even contradict these limitations:

> Jensen (1969), after recognizing explicitly that the heritability of individual differences within a population cannot validly be used as a measure of the heritability of the population means [of different races or classes], tries to do just that.[2]

Jensen seems to try to justify such contradictions as part of the creative effort of a pioneer in a field:

> Though I always heeded expert advice on purely factual and technical matters, I usually kept my own counsel on matters of interpretation and judgment. . . . There are always differences among investigators working on the frontiers of a field. They differ in their weighing of items of evidence, in the range of facts in which an underlying consistency is perceived, in the degree of caution with which they will try to avoid possible criticisms of their opinions, and in the thinness of the ice upon which they are willing to skate in hopes of glimpsing seemingly remote phenomena and relationships among lines of evidence which might otherwise go unnoticed as grist for new hypotheses and further investigations.[3]

Another experiment will bring out more clearly the peculiar "relativity" of the concept of heritability, as well as the thinness of the ice on which Jensen is skating when he uses this concept to boost IQ theory. Suppose that the seed is taken from a sack of an "open pollinated" variety of corn. Here there is a great deal of genetic variation. This time, however, we plant quantities of this seed in two carefully controlled environments, each one containing within it chemically identical soil and nutrients. But environment B has exactly half of the amount of nitrates as in environment A. In addition, zinc, an important trace element, is removed from the soil of environment B. Because the soil in environment A is perfectly uniform (or as close to uniformity as is possible) all of the variations we observe when the corn reaches maturity will be due to the genetic variations between the corn. The same is true of the variations we observe in environment B. Applying our heritability formula, g/g+e (where g is the genetic variance and e is the environmental variance) we arrive at a heritability of the corn in environment A which is 100%. Could we conclude from this something about the characteristics of the corn from the same batch in environment B? According to the common sense, metaphysical interpretation, 100% heritability means that all of the characteristic (such as height) is determined by heredity, nothing by environment. It would seem to follow from this that wherever we plant such highly heritable corn we would get the same results. The property should be entirely determined and fixed by the genes. In fact, of course, the corn in environment B would be quite different from the corn in environment A, despite the fact that in both cases the heritability of the corn is 100%. Paradoxically, this time the cause of the difference between the average heights observed in the two environments is entirely environmental.[4]

From this example it is clear that the heritability of a trait is *relative* to a given environment. It is not an estimate of the portion of that trait that is due to heredity alone, as opposed to environment. The *metaphysical* distinction between heredity and environment, which conceives of these as separate and juxtaposed entities, only wreaks havoc with our attempt to understand heritability. Since a heritability estimate is relative to a given environment it says nothing about what will become of the same individuals under environmental conditions different from the one(s) in which heritability has been estimated; it says nothing about the possibilities of change under other circumstances.

More fundamentally, in our opinion, it is important to understand that heritability tells us nothing about the particular characteristic *itself* under any circumstance. A heritability figure tells us only something about the *variations* that occur of that particular trait in a "population"; it tells us only to what extent the variations are due to genetic variations (and even this is done in a very limited way, as will be clear later). The confusion that easily arises in attempting to understand what such a statement means stems from a natural tendency to treat a variation or a difference as though it were a definite entity. This illusion results in part from the fact that differences are expressed in terms of definite numerical quantities. Heritability theory, including the technical method for establishing variance, attempts to treat variation as something in itself, irrespective of the real features of the thing that varies. Thus, in our previous example, 100% heritability told us nothing about the quality of our two samples of corn. The fact that we could measure the real height of the corn in each group enabled us to note that one group of corn was taller than another. This observation rested on the fact that we could tell how tall each stalk of corn was in itself. Heritability analysis, however, tells us nothing about the real features of the characteristic we are measuring. It deals strictly with observed differences, measured in relation to the average. Where there is no variation in the particular trait, even though this be genetically necessary, the heritability of the trait would be zero. (Consider, for example, the heritability of "one-headedness" among humans. Since there is no variation of this trait in a population, its "heritability" is 0, although it is completely caused by "heredity.") This abstraction of variation as a seemingly independent reality is something that heritability methods have in common both with IQ and with statistical "intercorrelations."

## Jensen's "sociology"

A final look at our second experiment will bring out the degree of caution that must be exercised in determining heritability estimates under conditions which are not absolutely controlled. In his article from which this example is borrowed,[5] Richard Lewontin asks us to suppose that the composition of the soil in environment B is not known to the investigator. What is observed is a lower average height in the corn grown in this soil. Upon analysis of the composition of the soil, the experimenter discovers the absence of nitrates, but fails to detect the missing zinc trace. He then repeats the experiment after doubling the amount of nitrates in soil B, and finds that, although the corn now reaches a higher average growth, it is still not equal to the corn in soil A. Where a Jensen might skate out on thin ice and proclaim that the cause of the differences is probably genetic, a closer study would reveal the inaccuracy of this conclusion.

This example is meant to show how much caution must be applied to the far more complex and less perfectly understood case of human development. But here we do not want to plead ignorance. Jensen makes due acknowledgment of possible future discoveries that might explain Black-white IQ differences environmentally. However, where he pretends to take account of environmental differences to explain IQ performance, taking differences in gross socio-economic status into account, he fails to consider an element that is known to everyone—the four hundred year history of racism and the special oppression of Black people in the U.S. It is not a matter of a scientist who is ignorant of a difficult to detect trace element, but of an assumption that a flagrant and absolutely central fact of U.S. history can be treated as negligible in explaining differences in IQ or academic performance.

Heritability analysis as applied to human development assumes that it is possible to equalize environment by the use of crude socio-economic indicators. Thus, if all the individuals in a particular school come from roughly the same kind of working class environment, the differences in intellectual performance would be "explained" by the genetic factor that is left once we have crudely dismissed the environment as an explanation of differences on the grounds that all of the families have roughly the same income, educational background, housing facilities, etc. This concept is stretched to the limit in the meritocracy theory which assumes that since environmental opportunities become equalized for everyone in the U.S., the observed differences are increasingly due to genetic differences. The leap from our chemically controlled soil for the growth of corn to equality of opportunity for the growth of our children may leave some of us gasping for air.

We wonder what sociologists Jensen has been reading in his effort to explain differences in IQ who omit among their "socio-economic indices" the special character of racism against Blacks in a country that was built in large part on the slave exploitation of Black people who were both systematically terrorized and consciously kept in a state of illiteracy. And enormous superprofits are still made through the substantial wage differential that continues to be maintained from doubly and triply exploited Black and other minority workers. A study of the history of the struggle of Black Americans shows that fear and ignorance were the tools of their brutal domination—not natural attributes of an inferior race. Roy Brown thinks that Jensen's "tentative conclusion" that

'Heredity . . . plays some role in the heavy representation of Negroes in America's lower socio-economic groups' [is] unbelievable when one considers the fact that absolutely nothing is said about the extreme deprivation that blacks have endured—300 years of the cruellest slavery known to mankind: 100 years of barbaric servitude, murder, lynching, burning, and intimidation, superimposed with an arrogant savage con game. There was literally no intention of treating blacks as human beings, but, rather, they were to be exploited and kept in servitude by any and all means, legal and illegal.[6]

Lack of real consideration of this central fact of U.S. history, as well as the deep-rooted, inhumane and devastating character of continuing racial oppression, shows how feeble has been Jensen's attempt to find "environmental" explanations for IQ differences—understood not as indicators of innate capacity but as indirect signs of suppressed abilities. The fundamental role that racism plays in the United States cannot be reduced, in vulgar materialist fashion, to "socio-economic indicators" that lump together social and national groups with qualitatively different historical characteristics. Despite equally savage treatment, the history of Native Americans cannot be equated with that of Blacks, to say nothing of Taiwanese immigrants (whose higher scores and low socio-economic status gives Jensen cause for further genetic hypothesizing). The national character and the forms of oppression of the peoples mentioned by Jensen cannot be reduced to a scale of gross living standards, however significant these might be.

Historical materialism is not disproved by the lack of full explanatory value of the crude sociological categories with which Jensen identifies the "environmentalist" position. Marx in fact explicitly criticized the mechanistic environmentalist theories of education developed by enlightenment and utopian socialist thinkers who held that "men are products of circumstances" and forget or could not yet recognize that "circumstances are changed precisely by men and that the educator must himself be educated."[7] We have already seen that the concept of an environment as something external to the people who make it stems from the real separation of individuals from the historical conditions of life and thought.

It is based on a system which "reconnects" the individual with the necessary instruments of human existence only on the condition that this expand capital at the going rate. The limitations of liberal "enlightenment," at the ideological basis of programs such as Headstart, stems in part from a failure to recognize the extent to which economic exploitation and superexploitation continue to stifle the development of the working population, white and Black. But far from drawing anti-enlightenment or anti-environmentalist conclusions from such limitations, Marx deepened the concept of the environment and the materialist theory of knowledge by showing that education is intimately linked with the self-development of the majority of the people struggling for full control over the conditions of their physical and intellectual existence. It was just such a struggle that led to the reforms embodied in the Headstart program, and only the enlargement of such struggle will stop the growing assault on the educational rights of children and adults, both white and Black.

## Heritability and widthability

A further example should help to clarify more precisely the limited and essentially pragmatic meaning of heritability, and the inability of this concept and its methods to deal with the real interaction of heredity and environment. This is true both on the strictly biological level and in the broader question of the relation between biological laws of evolution and the specifically social laws of history.

An example from the domain of geometry, reproduced by Howard Topoff (1974)[8], makes the same point that Lewontin does regarding the heritability of corn. There are particular features of this example, however, which enable us to question more closely the assumptions of heritability estimates. Suppose that there are two sets of football fields, A and B. The lengths of all of the fields of set A are 100 feet while all of those in set B are 50 feet. The widths vary in each set, however. In the first, the widths are 48, 49, 50, 51, and 52 feet; in the second, they are 23, 24, 25, 26, and 27 feet.

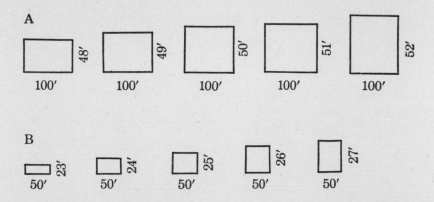

To determine the "variance" of the width, we find the mean or average width in each set, subtract each individual width from the mean, square these remainders, add them together, and divide by the number of cases. In both sets we arrive at a variance of 2. The variance of the length is, of course, 0—since the length does not vary. If we follow the "additive" model for heritability to determine the "widthability" (the ratio of the variance of the width over the total variance) we will use the following formula: $W = V_W/V1 + V_W$ ($V1$ = Variance of the length and $V_W$ = variance of the width, while $W$ = "widthability.") Accordingly $W = 2/0 + 2 = 1.0$. Since there is no variation in the length of any field, the widthability in both sets is 100%.

Variance $= \dfrac{\text{Sum of (Mean-Width)}^2}{\text{Number of cases}}$

Formula #1  $W = \dfrac{V_W}{V_W + V1}$    $W_A = 1$    $W_B = 1$

Formula #2  $W = \dfrac{V_W}{V\,(1 \times w)}$    $W_A = .0001$    $W_B = .00158$

Formula #3  $W = \dfrac{V_W}{V\,[2\,(1 + w)]}$    $W_A = .25$    $W_B = .25$

This example is likely to be somewhat disappointing since although it makes the same point as in the example of the heritability of corn, it seems intuitively to be even more trivial. This is true not only because variance in width, as with variance in the height of corn, gives no indication as to the fact of the important and real differences between sets of objects, whether corn or rectangles, but because the general formula itself appears to be completely arbitrary.

What reason is there to define the "total variance" as the sum of the variance of the length and the variance of the width, especially if this is supposed to represent the variance of the "phenotype" or observed phenomenon? In fact, the *real* "phenotypic variance" is the variance of the *areas* of the sets of fields, which has not been calculated. To determine the variance of the areas, however, it is not only necessary to have the *variance* of the widths and the *variance* of the lengths, it is necessary to have the "absolute measure" of the lengths and widths themselves. We need to multiply length with width in each case, determine the mean area, square the differences from the mean, add them up and divide by the number of cases. The variance of the area in set A is 20,000 sq. ft., and in set B it is 5,000 sq. ft. Assuming now that "total variance" means variance of the areas, our widthability formula changes considerably. $W = V_W/V(1 \times w) =$ 2/20,000 or .0001 for set A, and 2/1270 or .00158 for group B.

The same operation could be made where the "phenotype" variance was considered to be the variance in the *perimeters* of the fields. Our formula for W in that case would be $W = V_W/V2(1 + w)$. In this case we might double the number of widths, since two widths in fact vary in each case. However, since we should also probably have to double the number of cases in the denominator, $V_w$ would continue to be 2 in both sets. The variance in the denominator would be the same in both sets, this time, however, giving a figure for W in each case of .25.

What is the significance of these calculations? In the first place they call into question the formula for heritability, especially the formula for "total variance," suggesting that this formula cannot be adequately represented by our knowing only the *variance* of each factor. In addition, we must know the actual dimensions of each factor on an "absolute" scale, and the type of *relations* which these factors have with each other. In the determination of both area and perimeter these figures are known; in the first dimension the factors have a multiplicative, and in the second an additive relation to each other.

Suppose that we knew neither how long each side was, nor what relation each had to the other. All that we were to know was that the width had a variance of 2 and the length had a variance of 0. This is the situation that we find in Jensen's use of population genetics, where we know only variances or "deviation scores" but not absolute values. Jensen writes that heritability deals with "differences among individuals and not with some absolute amounts of some attribute" and that "an absolute scale, though preferable for certain purposes, is non-essential for heritability analyses so long as we think of the phenotype values merely as deviation scores. Nearly all psychological test scores are only deviation scores."[9]

Knowing only the variance of the length and the variance of the width, we in fact know little about the variance of the phenotype, except that it varies in some unknown way not totally unrelated to the variance of the width. We may know that since only the width varies, the "phenotypic" (area or perimeter) variance must be a function of that variance, and not of a variance of the length. We would be entirely wrong, however, to define the phenotypic variance as merely the sum of the variance of the length and the variance of the width. This definition is only a result of the poverty of our knowledge—for all we know, by our hypothesis, is that $V_W=2$ and $V_1=0$. It would be disastrous to multiply these figures or divide them, since the result would be zero and we would end up with nothing at all. It is unlikely that we should think of subtracting the figures since we intuit that they have some kind of "positive" relation to each other. Jensen suggests a functional model, P (phenotypic variance) = $f(G, E)$ (is a function of the variance of the genotype and the variance of the environment). He says that this is potentially valid ("like all models") but is not simple, and besides, the additive model is verified in agricultural genetics.[10] We will return to agricultural genetics. But for the moment it is clear that the functional model is inadequate for perimeter or area as long as we do not know what the functional relation is and what the absolute quantities are whose variances alone are known to us. Thus, in the absence of any other information, we would have to make the best of the formula for phenotypic variance as given in heritability estimates. This may in fact be useful in the practical conditions of agricultural breeding, but then it should not be improperly stated as a definition of the variance of the phenotype, suggesting a deeper theoretical understanding than is in fact warranted.

## How much or how?

Jensen devotes a section of his book *Educability and Group Differences* to a criticism of "Heredity, Environment and the question 'How'" by Ann Anastasi. W. F. Overton[11] bases much of his criticism of the "additive model" in the heritability formula on Anastasi, arguing that explanations in biology have moved from questions dealing with "which" factor was determining or "how much" of each, heredity and environment, produced the phenotype, to the question "how" heredity and environment interact to produce the phenotype. Overton argues that between the first two questions and the third there is a fundamental difference in general conceptual frameworks—the additive model versus the "interactive" model.

In defending "additivity," Jensen argues that the developed formula for phenotypic variance includes any interactive or "multiplicative" effect. The elaborated formula for phenotypic variance, according to Jensen, is $P = G + E + GE$, where "GE represents $G \times E$, i.e., the interactive or multiplicative effects of genetic and environmental factors."[12] Does Jensen's GE account for the "multiplicative effect" as shown in our example of area variance? A comparable effect in that example to what Jensen means by $G \times E$ would occur were the variance of the width to produce a change in the variance of the length. Of course this does not happen, and such a "multiplicative effect" is not the kind that takes place when a variation in the width produces a variance in the area because of the relation of the absolute width to the absolute length. Although Jensen writes of $G \times E$, he means that *variance* of G produces a change in the variance of E.

In fact there can be no "multiplicative" or *even* "additive" effect such as seen in our examples of area and perimeter variances unless we know the absolute measure for both factors and "how" they in fact interact. Knowing neither, Jensen is right in saying that the "simplest" thing to do is to add the genetic variance, the environmental variance, and any mixed variance where we can't readily distinguish the two. This would not *equal* the phenotypic variance (outside of simple quantities where such an addition would in fact produce the phenotype variance), but it would somehow affect it.

In this state of general ignorance we might hope that the little that we do know might have some practical usefulness. Jensen points to agricultural genetics as "verifying" the additive model. Thus if we know that the egg-laying capacity of chickens is highly heritable, "genetic selection rather than environmental manipulation is likely to yield the most rapid results" in changing the "phenotypic value of the trait."[13] But the "truth" of any pragmatically verified theory is limited to the conditions of the particular practice. The fact that a formula such as the above is "replicable" does not give it full theoretical validity. This is not achieved until we have a causal explanation of *how* the process takes place by knowing *what* the genetic activities are and *what* are the environmental circumstances that "interact" in the development of the trait. Were the egg-laying capacity of chickens 100% heritable, this only means that variations in this capacity have in the past not been affected by variations in nutrition given to the animals, amount of sunlight, musical entertainment, etc. It also means that good results have been attained by selectively breeding high egg-layers. This "high heritability" means that breeding has worked in the past, while nothing else has, so we should concentrate on that. Farmers knew this long before "heritability estimates" were used to represent it. Heritability figures may give a more precise estimate of how much attention should be devoted to breeding, based on the wider experience of many agriculturalists than an individual might estimate by intuition. Thus we do not mean to belittle the degree of value such estimates have, but only to show their purely empirical nature. Heritability is relative to a given form or range of environmental conditions. Egg-laying is highly heritable only given an established state of environmental conditions. And at any moment, an "environmental" change might prove vastly superior to any breeding method. Jensen is precise when he writes,

> The proportion of variance indicated by $1-h^2$ [i.e., the environmental component of the variance], if small, does in fact mean that the sources of environmental variance are skimpy under the conditions that prevailed in the population in which $h^2$ [heritability] was estimated. It means that the *already existing* variations in environmental (or instructional) conditions are not a potent source of phenotypic variance, so that making the best variations available to everyone will do relatively little to reduce individual differences. This is not to say that as yet undiscovered

(or possibly already discovered but as yet rarely used) environmental manipulations or forms of intervention in the learning or developmental process cannot, in principle, markedly reduce individual differences in a trait which under ordinary conditions has very high heritability.[14]

In both instances, however, we are dealing with a purely empirical procedure. That is, we cannot explain *how* the cure is achieved. We only know that herb "x" has been effective in the past, and it therefore makes sense to stick with it. Thus, when we learn how the genes controlling egg-laying in fact operate we may find that they do so by affecting the production of a certain hormone which leads to high egg-laying only when a certain combination of nutrients are assimilated. Artifically producing the hormone and/or providing an ideal diet may produce a much higher egg-laying capacity than could be achieved by the old selective breeding techniques. At this point "heritability" would drop for the chicken population in question, since environmental variation of hormones and nutrition now produce the main differences in egg-laying capacity. But instead of saying this, and inevitably implying the existence of a mysterious entity in the chickens, we could also say that the farmers have stopped concentrating on breeding techniques for producing high egg-laying chickens. The heritability of the chickens has become itself a function of the practical, technological, man-made environment of the chickens.

## Interaction, development and the metaphysics of variation

This detour into the heritability of corn and chickens and the "widthability" of football fields was necessary to clarify the precise and very limited meaning of the concept of heritability so that when we hear that "IQ is 80% heritable" we have a clear understanding of the meaning of this phrase—whether it is true or not. It has nothing to do with the real interaction of genes and environment—just as "widthability" has nothing to do with the real relations of the actual length and width or with the variations of the areas or perimeters. Heritability does not refer to the role that heredity plays as opposed to environment. It presupposes a given and unknown interaction of heredity and environment. For humans, it assumes a given interaction of biological laws of development and sociological laws whose real mode of interaction is unknown. Jensen himself makes all this clear, so as to gain the assent of the specialists in the field. Attacking "interactionists," who he charges are merely environmentalists in disguise, Jensen discards the argument that genes and environment interact and mutually influence each other in a complex way which cannot be gathered from heritability studies. In his 1969 article he dismisses this position as all-or-nothingism; i.e., either we must know exactly how the genes operate in the overall development of the organism, or we can say nothing about the relative importance of the genes. Jensen argues in 1973 that such "interactionist" critics do not understand the concept of interaction as used in population genetics (i.e., the "multiplicative" relation GE which we have examined above.) But their confusion is

> even more a failure to distinguish between (a) the *development* of the individual organism, on the one hand, and (b) *differences* among individuals in the population. To say that a growing organism, from the moment of conception, 'interacts' with its environment is a mere truism . . . and repeating the assertion that the individual is the result of 'the complex interaction of genetic and environmental factors' is simply stating the obvious. What the population geneticist actually wishes to know is what proportion of the *variation* in a particular trait among individuals is attributable to their genetic differences and what proportion is attributable to differences in their environment.[15]

These remarks explicitly confirm our detailed analysis of the abstract character of both IQ and heritability studies which say nothing about real historical development and attempt to treat variation as something in itself. Jensen's criticism of interactionists in biology is probably correct, and his "clarification" is accurate. But this is not to say that it is very meaningful. We have tried to analyze the limited and pragmatic meaning which heritability figures may have. Real biological scientists may be excused for assuming that heritability pretends to be a scientific concept and should be treated scientifically. Jensen takes for granted the "truism" that there is a complex interaction of genes and environment, and is not interested in a scientific analysis of the real process of development. However he is so immersed in the abstract world of "deviation scores" and of working in the dark with variations, without knowing anything about what it is that varies, that he seems to think that his "clarification" should settle matters, and that the distinction between development and variation and their separation has any reality other than in the metaphysical world of abstract concepts. And even there such abstractions continue to slip back and forth between the limited technical usage and the "pioneer" interpretations of Jensen. Carrying his rock of Gibraltar out onto the thin ice of metaphysical and idealistic interpretations, Jensen seems to think that it is perfectly natural to treat pure variations as definite realities with obvious meaning. It is this slipperiness of Jensen's footwork that enables him to indulge in technical refinements which disclaim that he is talking about "intelligence itself" or anything to do with real development, and then conclude that intelligence *in fact* does not develop. Throughout his work there is a "sleight-of-hand" in which the real questions of human development are "finessed," while abstractions, with the meagerest residue of reality clinging to them, are passed off as the real thing.

This sleight-of-hand is nowhere more obvious than where Jensen applies heritability to human education, seeing the heritability of intelligence as the basis for understanding "educability." Thus in a passage previously cited, Jensen makes all the necessary refinements for applying heritability to human education. High heritability means that *"already existing* variations in environmental (or instructional) conditions are not a potent source of phenotypic variance...." It is not clear what kind of instructional variations are meant here, but presumably he means that the actual programs to change the school system so as to improve the scholastic performance of low performers have not been very successful. Using the heritability model, since such environmental variations do not affect existing variations very much, these variations must be due to genetic variations. But how thorough-going have been the environmental variations? A horticulturalist may try to change the poor performances of some of his roses by giving them more water, when the basic problem comes from the clay soil they are growing in. Without a real knowledge of environment, how can one assert that the environment has actually been varied? The sprinkling of educational reforms to which Jensen refers are now drying up, with the help of his argument that we have tried and failed. Our gardener who stopped watering his roses because of similar thinking would soon lose his roses altogether. Thus if environmental variation is understood superficially, variations will be attributed to genetic differences which are in fact due to deeper structural differences in the environment which have been hardly affected by the "already existing" environmental changes.

Jensen guards himself against criticism on technical grounds stemming from the fact that heritability says nothing about changeability under different environmental circumstances: "This is not to say that as yet undiscovered (or *possibly already discovered but as yet rarely used)* environmental manipulations or forms of intervention in the learning or developmental process cannot, *in principle,* markedly reduce individual differences in a trait which under ordinary conditions has very high heritability." We have underscored Jensen's parenthetical remark and the words "in principle" (in other citations above the underlining was Jensen's own) because they point to the technical significance of the concept, while implying that *in practice* such marked changes will "probably" not come about. In any case, this assertion completely contradicts the main thrust of Jensen's work which is that environmental changes will never affect a trait so solidly embedded in the individual as "g." It shows how little the concept of heritability can theoretically support the idea of innate and fixed IQ. However, it constitutes one more "argument" in the accumulation of arguments which Jensen uses to make this concept plausible.

## Mendel and evolutionary interaction

H. Topoff, who defends an "interactionist" position and questions the validity of heritability studies as offering real scientific comprehension of the developmental process, carries his criticism back to Mendel's own experiments:

> Imagine for a moment that you were assisting Gregor Mendel as he conducted his pioneering crosses of inbred, homozygous garden pea plants. Mendel crossed plants having red flowers with those of the white variety; all of the offspring (the F1 generation) had red flowers. Mendel's conclusion was that red flower color is inherited. Suppose you had asked Mendel how the hereditary factors (which we now call genes) produce flower color during the development of any one individual plant from the seed to the mature organism. Mendel's reply could only have been, "I don't know."[16]

To say that a particular trait is inherited in this context is not to say that environmental factors played no role in its development. It is rather to say that the variation of *certain,* usually obvious, environmental conditions, does not affect the distribution of a trait in the given population. Thus, argues Topoff, Mendel raised his plants under equal conditions of light and water, and in spite of this "equal" environmental treatment, the color of his flowers showed pronounced variability—some being white, and some red. A deeper understanding of the complex conditions of development makes this concept of "environment" appear superficial. Although light and water are certainly the most pragmatically accessible environmental conditions, other factors from the very beginning of fertilization, in the seed itself, are external to the chromosomes carrying the genes and begin an interactive process, the knowledge of which in developmental genetics brings out the deep penetration of "environment" in the very functioning of the genes. Topoff writes that "The only valid conclusion from Mendel's experiment is that light and water were not responsible for the variability in flower color. Studies of heritability can only point out that certain factors are not responsible for the observed phenotypic variability."[17]

The fact that Jensen refers to his work as part of a "Mendelian revolution" makes it necessary to understand more precisely the "field of action" (to quote Engels) of Mendel's discoveries. It is often pointed out that Mendel's experiments were published in 1865, only six years after Darwin's *Origin of Species,* and remained basically unrecognized until 1900. This remark is sometimes made to suggest that evolutionary theory developed without any knowledge of genetics. The opposite notion also is true: genetic theory developed without any real connection with evolutionary theory. In fact, Mendel's theory of the genes as remaining unchanged throughout generations makes it essentially impossible to understand evolutionary change. It was only when it was understood how genes themselves could change that genetics could become reconciled with evolutionary theory. Their stability is only relative, not absolute, as was implied by Mendel's work.

There is a clear Mendelian bias in the concept of "heritability." In his breeding experiments, Mendel was interested in the *variations* that occurred in certain sharply contrasted traits of his pea plants. Thus while the first generation that resulted from the crossing of red-flowering plants with white-flowering plants consisted in all red-flowering plants, in the next generation, based on the self-fertilization of the previous generation, on the average one out of four plants was white, while the rest were red. Further self-fertilization of these plants showed that the white plants always bred white, and one-quarter of the red plants always bred red. The remaining fifty percent of the plants, which were red, produced red and white plants in the ratio of three to one—just as did the first generation reds. Mendel concluded that certain "unit characters" (the genes) determined flower color, and that these were of two types, one dominant and one recessive. These variants of the gene are called alleles, and are commonly symbolized as "A" for the dominant gene and "a" for the recessive. These two types of the gene that affects flower color in pea-plants combine in three possible combinations, according to the laws of chance. These "genotypes" are type AA, type aa, and the hybrid type Aa. The first two types will breed true, when self-fertilized, while the third will produce all three types again, in the proportion of AA, 2Aa, aa. Since the dominant gene determines flower color, both the pure dominant and the hybrid genotypes will produce red flowers. Thus the phenotype, red flowers, can be produced by one of two genotypes.

Essential to this analysis is the concept that variations in certain traits will occur in definite ratios, based on the assumption of two types of the gene that separate and recombine over the generations. When such ratios appear in fact and follow the genetic model, it can be said that all the variations are genetically produced and that flower-color is 100% heritable.

Suppose that after he carefully conducted his experiments and made his brilliant deductions, Mendel, dejected at the lack of receptivity to his ideas, allowed his garden to grow wild. When this garden was rediscovered thirty-five years later, suppose that most of the flowers were white, and only a few red ones could be found, mainly in inaccessible areas. This finding contradicts the ideal Mendelian ratio. After some search, it was discovered that the cause of this distortion was a particular species of bird that was attracted to the red-flowering plant, but left the white flowering ones alone. An environmental agent had intervened to "select" some of the plants for destruction. The actual variation of flower color among the pea plants in this particular area can be conceptualized as the product of genotypic variation *plus* some environmental selection that reduces the "ideal" heritability of the particular trait. "Total variation" can therefore be conceptualized as the result of these two factors. Note that low heritability here applies to a trait which is still "genetically determined." "Heritability" refers to the distribution of the trait in a population, not to its causes in the individual. Both white and red flowers are still produced by an interaction of genes and environment whose nature is not in question.

All of this of course leaves unanswered the question regarding "how" the genes operate so as to produce the particular color variations. The pea-plants of course grow up in a definite environment, which is essential to the formation of plant color. Gross variations of this environment, up to a certain limit, do not affect variation of plant color. Light, water, temperature, soil composition, etc., are all necessary for the growth of the plant and for its coloring process. *Variation* in color, however, does not depend on variation in these factors, and is linked with genotype differences.

The hypothetical bird that destroys red-colored plants is a *special* type of environmental intervention that appears to disturb the normal genetic-environmental equilibrium in which the environment factor is discounted only so far as providing an explanation of color variation. But this genetic-environmental equilibrium has its own history; pea-plants have not always existed. They evolved in the course of natural evolution through transformations of other species and under the decisive influence of environmental selection. In other words *general* environmental conditions are necessary for the evolution and normal growth of pea-plants. The independence of the genetic factors is therefore only relative, and has to be explained by previous natural history. Heritability theory abstracts from this "ancient history" and is concerned with how *special* environmental factors modify the variations that are predictable from knowledge of the genetic composition. The fact that "environment" is regarded as reducing heritability presupposes that ideal heritability is taken as a starting point. Such a conception, which distinguishes genotype effects under normal conditions of development and environmental alterations of a special kind, is valuable within a definite practical field of activity and analysis. Like most such antitheses, once they become absolutized and turned into "metaphysical" absolutes, they become barriers for further scientific development. Such a conception is applicable within *relatively* short time spans, during which the *special* environmental factor (e.g., the bird) can be conceived of as adding its effects to those of the genotype and detracting from the full expression of the genotype under *general* environmental conditions. From the point of view of the evolution of the species, however, the environmental "Darwinian" factor still stands out as the overall determining factor in the explanation of natural history. As a result of Mendel's discovery of the gene and further developments of genetic theory, evolutionary theory has been enriched and concretized, but not refuted.

The above examples of heritability presuppose either a definite knowledge of the genotype or of the environment as a basis for determining heritability. In many cases, probably in most instances of importance to agriculture, genetic structure is unknown or imperfectly known. The trait in question, such as milk productivity, weight, etc., is not the result of a single gene or a combination of relatively few genes. Where the trait results from the action of many genes it is said to be "polygenic" and estimates of heritability become more approximate. In ordinary breeding methods, the breeder selects above *average* qualities which may appear randomly in a given population. He then interbreeds the individuals so selected, in order to produce stock that only exhibit the desired properties. The new generation generally does not do as well as the selected parents, falling somewhere between the *average* of the original population and the *average* of the selected parents. Were the offspring to exactly repeat the characteristics of their parents, these characteristics would be considered 100% heritable. A simplified formula for determining heritability under these conditions is to determine the extent to which the selected parents are above average for the given population and then the extent to which their offspring is above the original average. Heritability is then defined as the ratio of the "gain in selection" of the offspring over the "selection differential" of their parents. Thus if the parents average twenty pounds above the average for their particular population and their children average only ten pounds above average, heritability of that trait is 10/20 or .50. Essentially it is argued that the variation of the trait in the original population is a result of genetically determined variation plus environmental variation. Since environmental conditions are kept similar for the selected parents and their offspring, the offspring are said to express the extent to which their parents are genetically above average. This explanation is a simplification of a more complicated situation, involving the possibility, e.g., that a combination of recessive genes in the parents may produce the relatively poorer showing of their offspring. Since this would be a genetic effect, its probability would increase the estimate of heritability. On the contrary, suspicion that the selection sample was getting special environmental treatment would diminish the heritability estimate. This entire experiment would of course be relative to a particular environmental context, which would be as controlled or equalized as possible. Heritability is seen to be still relative to environment. Furthermore, we are dealing with variations from the average or population means, and our heritability figure has to be evaluated as having meaning only in this context. Finally, we do not explain *how* the actual genetic structure produces the final result. Agricultural science has in fact made major strides precisely by leaning less on gross heritability estimates, and more on understanding the precise mechanisms of gene-environment interaction—where it is not possible to *juxtapose* heredity and environment as two separate factors. Far from being an empty truism, the concept of the gene-environment developmental process provides the key to major advances in agriculture. Heritability estimates can serve a useful, but limited function in this study.[18]

### The social environment and human activity

Thus even where breeding experiments produce marked phenotype variance, at most we can say is that the *immediate cause* of the variation is "hereditary"; but since this effect operates only in the complex gene-environment interaction, it is not possible to say that the ultimate or determining cause is "hereditary"—for breeding is always relative to a certain response to environmental conditions. Such a clarification is itself likely to be misleading if we were to apply it to human development, where, as we have seen, the polar relation of genes and environment, which determine the biological process, cannot provide a scientific basis for the study of real human development. Biological processes are integrated into a socio-economic process whose basic determining factors are the interaction of productive forces and relations of production in the unity-of-opposites of human society and nature. Even where criticisms are levied against the misuse of the concept of heritability, and the gene-environment developmental process is explained, it is still misleading to apply this "interaction" to the process of human development, or to suggest that IQ differences can be looked at in this framework. John Hambley writes of the "environmentalist" response to the "genetic" argument that "the tacit idea of manipulation of the environment to produce the 'appropriate' phenotype, while perhaps appropriate in agriculture, represents short-sighted paternalism in the human context."[19] More deeply, this approach implies a population that is only passively related to an "external" environment, and not the active determiner of its own development. The concept of an "enlightened" educator or class of technocrats who can mold the population according to an ideal preconception of life is the Skinnerian, behaviorist foil against which heriditarians construct their arguments that the real determinants of development are buried "within" the individual. Marx rejects this antithesis between a "contemplative materialist" theory of knowledge, on the one hand, and an activist, but idealistic theory of knowledge on the other. Thus in his first thesis on Feuerbach Marx writes:

> The chief defect of all hitherto existing materialism ... is that the thing, reality, sensuousness, is conceived only in the form of the *object, or of contemplation,* but not as *sensuous human activity, practice,* not subjectively. Hence, in contradistinction to materialism, the *active* side was developed abstractly by idealism—which of course, does not know real, sensuous activity as such.[20]

The debate that Jensen carries on with "egalitarian environmentalists" reflects a similar philosophical antithesis. Whereas the "environmentalists"—generally, behaviorists—argue that learning is the result primarily of external environmental inputs, Jensen argues that their programs have failed to produce any significant variation in the results of their educational efforts. The cause of variation, he argues, should therefore be sought in the "inner," i.e., genetic, part of the individual. What is missing from both analyses, to follow Marx, is social activity, human "self-changing," which today has the main feature of overcoming the historically developed and transient mode of separating individuals from their underlying and necessary social relations and from the main material and intellectual instruments of existence. To confine the argument to deciding between two sides of the separation between the individual and the "environment" is merely to reflect the *effects* of the underlying social structure, and to build speculative bridges from one side to the other. The real connection between the individual and his environment, involving a theoretically more essential understanding of the "environment," consists in one's practical, social activity which is increasingly impeded by the transient mode of production in which the means of life are privately appropriated even while they become more and more directly social. A deeper understanding of the real connection between the individual and the environment as social practice, Marx writes, ultimately has a revolutionary direction.

# 10

## Twins and other relations

*T*he main argument for the high heritability of human intelligence stems from the study of identical twins who have been raised in different environments. We have seen that arguments deducing heritability from sociological methods of equalizing environments are not very convincing. There are other studies, however, which show a high correlation between the IQ scores of individuals and their known genetic relationship. In other words, the more similar the individuals are genetically, the more likely will their IQ scores also be similar. Jensen presents an imposing table of "Correlations for Intellectual Ability: Obtained and Theoretical Values."[1] The "theoretical values" are estimates of the degree of similarity which would be predictable from a knowledge of their genetic relationships. The obtained value is the average (median) correlation of their IQs. Thus the theoretical value of unrelated children is .00, and the IQ correlation of unrelated children is also zero. On the other hand, unrelated children reared together have an IQ correlation of .24—indicating, says Jensen, the extent to which common environment affects intelligence. Siblings reared together have an "expected" or "theoretical" IQ of .52 (as estimated from a genetic model), while their obtained scores correlate .55 on IQ tests. Siblings reared apart have the same theoretical value as siblings reared together, since adoption does not change their genetic structure, while the obtained correlation is .47. It should be noted that the correlation of siblings reared together is not surprising, since they share a common environment—but their IQs are said to be much closer than those of unrelated children reared together (as in orphanages or adopted children), while siblings reared apart do little worse than those raised together.

The main genetic experiment, however, deals with the study of monozygotic twins, twins who have developed from a single fertilized egg, and therefore are genetically identical. Dizygotic "fraternal" twins, according to Jensen's table, have the same "theoretical" value as ordinary siblings, and their obtained scores are about the same. Since monozygotic twins are genetically identical, they have a theoretical value of 1.00. The obtained IQ correlation of such twins living together is .87, while the correlation of such twins reared apart is .75. It is this latter figure that is especially impressive, and is the main basis for the argument that the heritability of IQ scores is about .80.

It should be noted that the "correlation coefficients" given above are arrived at by certain mathematical techniques and have definite properties in relation to other functions of the statistical system.[2] Thus, if the IQs of siblings have a correlation coefficient of .50, this means that if we know the "deviation score" of the one child we can predict the deviation score for the other by multiplying the first by .50. Thus a child with an IQ of 115 is about 1 standard deviation from the average. That child's sibling will likely have an IQ of 1 × .50 or a + .5 deviation. It is necessary to translate this back into IQ scores, to get a score about half-way between 100 and 115. (This is an expression of "regression to the mean"—a statistical necessity for imperfectly correlated variables.) Another function of the correlation coefficient is to obtain the degree of "covariance" between two variables. This is obtained by squaring the correlation figure. This to say that if genetic variation has a .50 correlation with IQ variation there is a mutual covariance of .5 × .5 or .25. Assuming that covariance is roughly equivalent to causation, we can say that the 25% of the variation of intelligence found in the siblings is due to their genetic similarity. Thus if one child has an IQ of 115 and the other an IQ of 107 the argument is that 25% of this similarity is a result of their genetic similarity. Jensen "corrects" the estimate for identical twins reared apart, bringing the correlation up from .75 to .90. To get the heritability of IQ—the extent to which IQ variation is due to genetic variation—it is only necessary to square the .90 figure. This gives the well-known 80% heritability. One who was content with the "uncorrected" figure of .75 would only get a 56% heritability.

## *Is human heritability possible?*

Before examining the data on which such evaluations were based it is important to have a clear idea as to what such figures in fact mean. We will focus on the most striking "genetic experiment" which we find in the literature of IQ heritability, i.e., the case of identical twins raised in *separated* environments. Since the genes of such twins are identical it is argued that any diversity in their IQ scores will be due to the diversity of the environments in which they are raised, not to their genes. If the environmental diversity is great, but the IQ differences are small, then it seems that genetic factors are more "potent" than environmental factors. To be an accurate basis for estimating the heritability of a population, the separation of the twins must match the environmental differences in the population as a whole. Moreover, the resulting heritability figure is strictly relative to that population. We have to be clear about the meaning of this relativity, referring to our previous analyses of both heritability and IQ.

With respect to the relativity of heritability, we have seen 1) that heritability does not distinguish the contributions of heredity as opposed to environment in the overall development of the trait (IQ); 2) that heritability cannot be used to infer characteristics of the population under different environmental circumstances; and 3) that heritability figures for a given population cannot be used to explain the cause of differences between populations (e.g., between Black and white).[3]

When we relate these conclusions to our previous analysis of IQ we must recall our argument that IQ disguises qualitative intellectual development. It is a reflection of the consistency of the rank order of an individual at different stages of his or her intellectual development. We further argued that the qualitative development cannot be explained by some abstract "general intelligence," and that the variations at each level of development that is being analyzed are relative to the respective forms of development. They cannot be entirely explained, therefore, solely by biological differences between individuals. This does not mean that in explaining these variations biological variations may not play some role. This role, we have argued, is all the greater, as the level of intellectual and specifically human development is the lower. Thus, Engels hypothesized that variations in breeding methods under the influence of natural selection played a major role in the evolution of forms of the family in primitive societies, and this in turn favored the development of the productive forces, which was still the decisive factor in the physiological differentiation of human society from the higher animals. We suggest that a similar process is repeated in the development of individual human beings—although, as was true in the case of primitive human development, the social, human element remains decisive.

The notion that biologically related differences are more powerful to the extent that the performances affected are "primitive" is borne out particularly in the case of sex differences and patterns of relations between the sexes. Sexual maturity occurs relatively late in the physiological development of the child, at a time when the child or young adult is assimilating advanced elements of his or her culture. That patterns of sexual behavior have considerably varied historically is a well-known fact of anthropology, and is at the center of Engels' study of the origin of the family characterized by paternal domination.[4] Genetically related variations in sex play an important role in the life of the individuals of any society. But how these variations are *actualized* historically depends on their social "interpretation"—rooted in the level of development of the productive forces of society and the mode of appropriation (property) of these productive forces. This is a case in which a variation that has clear genetic basis, and is also necessary for the survival of the species (unlike relatively superficial differences in skin-color, etc.), nevertheless is transformed and infused with a social-historical content.

We might imagine a "strength quotient" (SQ) test which attempted to determine the heritability of strength—the extent to which variations in strength at each particular age correlated with variations in sex. Suppose that the test measured the performance of sixteen year olds at lifting bar bells. It was then discovered that "strength" correlated .90 with sex—the males being of course superior. Does this prove innate differences in strength? While physical strength appears to be an easily measurable quantity, in fact this impression is misleading. It is an illusion and an abstraction stemming from our habit of looking at human beings in a state in which they are reduced as much as possible to their immediate physical existence and deprived of the social and material instruments of life which in fact differentiate human beings from animals. Were human beings to have relied on such brute strength in order to exert an influence on the natural environment we probably would still inhabit trees. Real *human* physical strength involves the development of tools and forms of social organizations that primarily determine the strength of the individual in real life. Sex related differences in brute strength may in fact play some role in social life—all the more so as the productive forces of society are undeveloped or the form of activity is relatively primitive. Modern machinery, however, qualitatively increases real human strength and at the same time in practice tends to cancel out whatever physical differences in "brute" strength there may be "innately" between the biological sexes. Also, our "strength quotient" is relative to the population in which it is measured. Many peasant women may be stronger than many urban men in the U.S. today, but the IQ of methodology would normally miss this kind of comparison.

The fact that "true IQ" which correlates highly with both social class and ultimate success in school is only "formed" between the ages of four and six, seems to confirm the notion that biological differences have the greater impact to the extent that the activities involved are relatively primitive. We have argued that the truly social forms of development of intelligence are based on the historical division of labor in class society. While such class-based forces would be active from the very beginning of the child's life, their effect would become all the more pronounced as the child reaches the more complex, developed states of intelligence. At this point social-class relations including racial discrimination begin to dominate in determining the relative standing of the children in given age groups in the population.[5]

## Twin studies

"Identical" twins resulting from a single fertilized egg (monozygotic) have essentially the same genotype. Differences in their performances will therefore be due to differences external to this original genetic identity. Such differences begin with the separation of the original egg, and the developmental process that occurs in the womb. (Jensen believes that the main source of later IQ differences can be traced back to differences in the womb.)[6] Since the heritability is relative to a given population in a given environment we can have heritability estimates for activities within the family, within the social class or within the population of the nation. Since such twins are genetically identical, any difference between them would reflect the effects of environmental differences in the family. However, it would be difficult to say that such differences reflected the full impact of differences possible within a family since identical twins usually form a special bond or are "twinned" in such a way as to accentuate their similarities. Their environments are more "identical" than would be the case for ordinary siblings or fraternal twins. Genetic and environmental relationships are "confounded," and it is not possible to argue that, because twins are much more alike in behavior (or IQ) than their brothers and sisters, this is the result of their genetic similarity. Whatever the effect of this genetic similarity on their development, it is soon given a social or "environmental" form—whether because of "twinning" or because of the twins' own gravitation toward each other and special social unity.[7]

It is studies of twins *raised apart* that claim to give a convincing estimate of the heritability of IQ. If twins are raised in different environments and still consistently score in a highly correlated fashion, then it might be concluded that genes play an important role in the development of IQ or intelligence. Here it becomes a question of how different the environments are. High correlation of children who grow up in similar neighborhoods with roughly similar environments will reflect a combination of the similarity both of genetically related processes and of environments. If their scores are more highly correlated than those of the average children reared in similar circumstances, then such results would give a "within class" heritability. This is to say that children with identical genetic structures will tend to develop in a more similar manner than genetically different children raised in a similar environment. It seems obvious, however, that such an effect is relative to the environment. Were identical twins raised respectively among the Hopi Indians and the Syrian Bedouins we would expect them to differ in graphic abilities about as much as the average children for those societies. We would expect to find enormous differences between identical twins raised respectively in a stone-age tribe and in a U.S.

suburb. "Heritability" is therefore relative to the extent to which environment is varied.

Since we do not have studies relating to such considerable separations, we must limit ourselves to studies of identical twins raised in the same country. Jensen's claim is that IQ is 80% heritable within the U.S. population for white native-born Americans. This means that 80% of the differences of IQ in this population are due to differences in the genotypes "determining" intelligence. Thus, if one twin is raised in a family at the lower end of the socio-economic scale, which normally produces the lowest IQs, it would be found on the average that the other twin, raised in a professional family which normally produces the highest IQs, would have an IQ score correlating .90 with the first twin.

Such a finding we believe would be very significant—though it would not have the meaning that is normally given to it. We want to ask now whether such results in fact are true.

## Kamin's analysis of the data

The main argument is that heritability is high for the "population as a whole." That is, identical twins raised in widely different social classes will nevertheless have similar scores on an IQ test. This claim has been based on Jensen's appraisal of the only four such studies in the IQ literature. These studies are meticulously analyzed by Leon J. Kamin in *The Science and Politics of IQ*. The value of Kamin's work in refuting the claim that the heritability of IQ is substantial for the population receives indirect testimony from Jensen himself.

In 1969, Jensen, who regarded Burt's study of kinship correlation as "the most satisfactory," was especially impressed with Burt's studies of separated twins because they were based on the largest sample (reaching over a period of years a total of 53 pairs), because the average IQ and the standard deviation were close to the population norm, because all the twins were separated within the first six months of their life, and "most important" because "the separated twins were spread over the entire range of socio-economic levels."[8]

Far from taking Burt at his word Kamin carefully studied Burt's work. He showed the lack of scientific method, even within the terms of IQ theory, serious contradictions, as well as astonishing consistencies—such as the fact that between 1955, when Burt measured the IQs of only 21 pairs and 1966, with 53 pairs, the IQ correlation remained identical to the third decimal point (.771)—a remarkable testimony, Kamin ironically remarks, to the power of heritability. Since both Jensen and Herrnstein assert that Burt's was the only study which was based on twins raised in completely "uncorrelated" environments, Kamin was especially interested in the evidence for this. But Burt gave no positive descriptions other than his word that children came from one of six classes from "higher professional" to "unskilled." Kamin had attempted to analyze the statistical results themselves. There he noted statistically unusual features which belied Jensen's claim that the variance among the population was normal. In fact, a more detailed analysis showed that there were peculiarities both in the scores and in the variances of the children raised in upper-class foster homes as well as their twins. Thus, twins raised in upper-class foster homes had IQs below those raised in lower-class foster homes. The variance of the upper-class foster children was only 7.61 points, well below the average population variance of 15 points; their twins also had little variance between them. Kamin writes that "There appears to be a force at work which causes children reared in upper-class foster homes to have relatively low IQs, with remarkably little variance."[9] He concludes that this force was probably not the genes but Burt himself who provided almost no evidence as to how he produced his assessments of either IQ or environment. In fact, while Jensen asserts that Burt used an English version of the Stanford-Binet, and followed the rigorous standardized procedures which psychometricians regard so highly, Kamin shows that Burt relied more on his own intuition and presuppositions than anything else. Driven back to the original source in order to refute Kamin's critique (which was originally circulated in an unpublished paper), Jensen writes that " . . . alas, nothing remained of Burt's possessions . . . unfortunately, the original data are lost, and all that remains are the results of the statistical analyses. . . . "[10]

Kamin concludes:

> We have seen that Burt's data, reporting by far the strongest hereditarian effects, are riddled with arithmetical inconsistencies and verbal contradictions. The few descriptions of how the

data were collected are mutually inconsistent, as are the descriptions of the "tests" employed. The assessments of IQ are tainted with subjectivity. The utter failure to provide information about procedural detail can only be described as cavalier. There can be no science that accepts such data as its base.[11]

Other studies which have in fact produced data on the procedure for obtaining IQs and environmental assessments show that most of the so-called separated twins were raised in relatively similar environments. The Shields study of 37 pairs of separated identical twins provided more detailed descriptions of the environments of the "separated" twins. Twenty-seven pairs were raised in related branches of the same family. The scores of those raised in related families was correlated .83, while those in the "unrelated" families was only .51.[12] Most of the children raised in the "unrelated" families, moreover, were raised by family friends. One example of a pair included in the *unrelated* family category was Jessie and Winifred, separated at three months: "Brought up within a few hundred yards of one another. . . . Told they were twins after girls discovered it for themselves, having gravitated to one another at school at the age of 5. . . . They play together quite a lot. . . . Jessie often goes to tea with Winifred. . . . They were never apart, wanted to sit at the same desk. . . . "[13]

The Newman, Freeman and Holzinger Study of 19 pairs of twins, done during the depression in 1937, introduced a systematic bias into its selection sample (who were brought to Chicago at great expense and rewarded for being "right" for the study) by explicitly rejecting applicants who said they were unlike. Thus one pair of twins was refused the trip to Chicago because one of them wrote that while they were often mistaken for each other they were "as different as can be in disposition."[14]

The Juel-Nielsen study of only nine twins was found to be equally unconvincing regarding its procedures (use of an unstandardized test) and its assessment of environment. Thus in their summary, the authors write of Peter and Palle that "the standard of the homes in different parts of Copenhagen were socially very different." But in the description of these twins earlier it is said that "Their homes do not seem to have differed particularly as regards social and economic status, housing conditions or general cultural influences, but their childhood environments were otherwise very different."[15] Whatever this last statement might mean, this final study again produces no evidence of any heritability *for the population as a whole*. We can only conclude that whatever effect genetic similarity may have for IQs of "separated" twins, this effect is relative to a narrow environmental range of variations and the effect is nowhere shown to survive real inter-class variation.

## Kinship correlations and adopted children

When we look at the data for other kinship correlations we seem to be on no firmer grounds. We will not summarize Kamin's detailed study of the Erlenmeyer-Kimling and Jarvik (EKJ) table of kinship correlations matching the expected correlation on "theoretical" grounds with the actually arrived at empirical correlations, which is another strong argument for high heritability. We have seen that the most striking correlations regarding separated identical twins do not provide grounds for substantial heritability in the population (even with all the qualifications which this concept requires when it is correctly understood). We only note a few striking points. In the first place, this table, said to summarize 30,000 correlation pairs and 52 separate studies, has been called "a paper that condensed in a few pages and one figure probably more information than any other publication in the history of psychology."[16] Kamin says that it would be difficult to underestimate the impact of this paper, which has been reproduced widely and is used by Jensen, Eysenck and Herrnstein. The studies on which it is based, which are generally said to reflect widely divergent views of the heredity issue, are nevertheless not listed, and it is impossible for the researcher to check up on them without writing to Professor Jarvik and asking for the list of studies which his paper is said to summarize. Kamin's chapter analyzes this original material and notes that the "divergences in viewpoint" are more like family quarrels among hereditarians.

One noteworthy fact is that both Jensen (1969) and Eysenck (1971) report as many as 33 studies on siblings raised apart resulting in a median obtained score of .47 against a theoretical score of .53. This, when matched with the fact that siblings raised together show an obtained median score of .55, seems to provide striking evidence for the hereditarian view. With tongue in cheek Kamin writes:

"There can have been few environmentalist readers whose dogmatism was so blind as to withstand empirical evidence of this quantity."[17] The interesting thing is that the EKJ paper mentions only *two* studies of this phenomenon, though the median score was .47 also. It was Professor Burt who came up with the conclusive 31 additional studies which he claimed came from "a number of studies, chiefly British, which do not appear in the American collections." Following in the footsteps of EKJ, Burt gave no references. Pressed hard from critics, Jensen questioned Burt on this, and reports in 1974 that "Burt gives the 'number of investigations' of siblings reared apart as 33; I questioned this to Burt and he said it was a misprint—the correct number is 3." Kamin writes: "There has been a recent recognition that something may be, if not rotten, at least inaccurate in the state of England."[18]

On closer inspection of the studies, it appears that one of the two entered in the EKJ paper was by Burt himself who investigated 131 pairs of siblings raised "apart"—giving a correlation of .44 for "group test," .46 for "individual test" and for "final assessment," involving Burt's own more insightful perception of the situation, a "corrected" figure of .52—exactly what Burt's theoretical model would predict were the genes to play the exclusive role in determining variations in intelligence. EKJ, however, entered the more modest figure of .46. This is made up for in relation to the Freeman, Hollzinger and Mitchell study in 1928 of 125 pairs of separated siblings. This study gives two figures, based on two methods of calculation—.25 and .34. EKJ entered the .34 along with Burt's .46 to give a median of .40. Another study, mentioned in the reference list sent to Professor Kamin by Dr. Jarvik was not entered into the computation of the median obtained score. For 78 pairs of separated siblings, Hildreth in 1925 found a correlation of only .23, a figure that would change the median from .40 to .25—a far cry from the .47 figure for no less than 33 studies which was reported by Jensen in his 1969 paper.

It should be noted that while this figure nevertheless shows a higher figure than the theoretical correlation of .00 for the public at large, a more relevant comparison would attempt to match individuals with comparable environmental backgrounds to those in the study. One such study, noted by Kamin but omitted from the EKJ summary, was done by Verner Sims at the University of Alabama in

1931. Sims was interested in the IQ correlations of siblings raised together, and found a correlation of .40 for 203 pairs, below the theoretical prediction of .52. At least partly aware of the relativity of such figures, Sims wrote: "Presumably children paired at random would show no resemblance, but one cannot compare paired siblings with random pairs and account for the differences in terms of inheritance. It is only when environmental differences are equal that comparisons have meaning."[19] Although it was not possible to find unrelated children raised in exactly the same environments Sims selected 203 pairs of unrelated children from the same school, matched as closely as possible to the true siblings for "home background" and age—using a rough social class questionnaire to determine "home background," the limitations of which Sims himself recognized. The result nevertheless was an impressive correlation of .29 for these unrelated pairs, as compared with a .40 for the related ones raised together.

Another study which is considered quite important in the "IQ heritability literature" is the 1949 study of 100 adopted children by Skodak and Skeels. This study reported a significant correlation (.44) between the IQs of the children and those of their biological mothers, with whom they had not lived. On the other hand, their IQs correlated an insignificant .02 with the adopting mothers. The children were given Stanford-Binet tests on four different occasions over a span of about eleven years. Since no IQs were determined for the adopting parents, the IQs of both biological and adopting parents were estimated from knowledge of their educational levels. While these conclusions were widely cited in the literature as evidence for the strong effect of heredity, Kamin notes that the previous 1945 paper had shown quite different results for 139 children. Then, the IQs of the children correlated .24 with their biological mothers' and .20 with their foster mothers'. Between 1945 and 1949, 39 families had dropped out of the study changing the nature of the sample significantly. Fifty-one per cent of the 100 foster mothers who remained in the study had attended college. As time went on, the group of foster mothers became more and more alike in terms of education. Those who had attended college tended to stay with the study to the end. What effect does this have on the disappearance of correlations between adoptive mothers and children?

Kamin makes a point that we have stressed throughout this

essay. Suppose, he argues, that a study is done to see whether a particular disease is caused by environmental or genetic factors. One study finds that the existence of the disease systematically increases with decrease in income. Another study finds that there is no correlation between income and the existence of the disease. Such a contradictory finding is compatible with two different ranges of income. While there may be high correlation between the occurrence of the disease and income where the range of incomes is between $1 and $7,000, there may be no correlation where the range is between $7,000 and $14,000. However where a study uses only the *abstract variance* in income, and does not relate this to the qualitative features of that income range, both samples would be considered identical; both would give the same variance. Kamin expresses this by saying that one must not only compare the variances but also the means. In other words, there has to be some knowledge of the phenomenon that varies, besides the fact that it varies. One must know what the average represents, beyond the fact that it is an average.

In relation to the study of Skodak and Skeels, this simply means that while something may correlate with education of parents who dropped out of school between grades four and eight, one should not expect the same thing to happen for parents who have between one and four years of college. Correlation means simply that as one variable increases another will increase in a systematic way. Thus the correlation between biological mother and child means only that the more education the mother had the higher was the IQ of the biological child. It does not mean that the biological mother's IQ was close to that of her child. In fact the children's IQs averaged far higher than their biological mother's. Thus there can be a "correlation" between IQs of mother and child where the less educated mother has a child with a lower IQ—even if this is far higher than his mother's—while the more educated mother has a child with a higher IQ than that of the first child. On the other hand, the fact that the adopting mother had 12 years of school or 16 did not correlate with any systematic difference between the IQs of their adopted children. The 12th grade mother did not tend to have children with lower IQs than the mother with four years of college.

On the other hand, the average IQ for the adopted children as a whole was 117, while the mean IQs for the 63 biological mothers who had been given IQ tests was 86. This is a jump of 31 IQ points. (This IQ comes from the final testing, based on the 1937 Stanford-Binet. Herrnstein, who tries to diminish the significance of this fact, begins by using the average score of 106–107 according to Kamin—which was obtained by testing the children on the 1916 version of the Stanford-Binet, which was the version their biological mothers were tested on.)[20] Thus, paradoxically, this study is used to show both heavy hereditarian influence and the opposite. In fact, a correct understanding of the use of "intercorrelations" shows the invalidity of the hereditarian interpretation.

As for the correlation between IQs of biological mothers and those of their children, Kamin argues that this can be explained by the selective practices of adopting agencies and the different type of agencies from which the children were obtained for the study. Adopting agencies do not randomly select children for adoption, but, perhaps believing that IQ is hereditary, attempt to match the adopting parents with the biological parents. Kamin also argues that parents who adopt children cannot readily be compared with ordinary parents. Comparisons which assume that their environments can be equated with those of natural parents and their children by use of socio-economic indicators probably miss important differentiating factors.

For further study of the data and the formulas on which other forms of heritability estimates are made, we refer the reader to the very interesting book by Kamin. The Kamin study also has valuable chapters on the history of the political uses of IQ tests and the racist and national-chauvinist ideologies which accompanied their early development and practical use in the United States. We have confined our attention primarily to an analysis of the theoretical concepts underlying the notions of IQ and heritability as interpreted by Jensenism and as widely broadcast in the public media. Professor Kamin's book, as well as a number of other books and articles critically examining IQ ideology, testify to the social consciousness and truly scientific spirit that exists in the academic world.

# 11

## Real science and real freedom

Jensen and others have added one further "argument" for the support of their position—the "argument from scientific openness." According to this argument, nothing in science is "absolute"—popular opinion in this matter to the contrary. As a result, someone with a truly scientific spirit should be "open" to various hypotheses, and bold pioneers like himself should be willing to skate out on thin ice where technical distinctions would make others more cautious. According to Jensen, "The nature of scientific 'proof' is poorly understood by most non-scientists. It is surely not an either-or affair, and in fact the term 'proof' is inappropriate in the empirical sciences."[1] We think that we have shown at least that the term is inapplicable to Jensen's own empirical studies. Jensen argues that real scientific progress is made by "weakening the explanatory power of one or more competing hypotheses and strengthening that of another on the basis of objective evidence."[2] This is a complex process, Jensen writes, which requires understanding the basic assumptions underlying a particular theory, the limitations of the theory, and the extent to which the theory explains the complexity of the data without adding *ad hoc* hypotheses. He concludes that as a result of such an approach "a largely genetic explanation" of class and racial differences in educational performance is "in a stronger position scientifically than those explanations which postulate the absence of any genetic differences in mental traits and ascribe all behavioral variation between groups to cultural differences, social discrimination, and inequalities of opportunity—a view that has

**173**

long been orthodox in the social sciences and in education."[3] He then goes on to accuse opponents who adhere to "egalitarian environmentalism" of "intellectual fastidiousness," of being guided by "sheer respectability," "preordained notions and inhibitions concerning what is and what is not respectable grist for research" and "research taboos."[4] In case anyone still has any doubts regarding the "largely genetic" explanation, Jensen assures them that "in complex subjects this is a gradual process punctuated by ambiguities and doubts, gaps and inconsistencies," but ultimately "as the work progresses a preponderance of evidence favors certain key hypotheses and leads to the abandonment of others."[5] Basically, a "no holds barred" approach is best for the truth, and best for minority groups as well "whose plight is now in the foreground of public attention."[6] Attitudes of those who would rule out any "reasonable hypothesis on purely ideological grounds" "constitute a danger to free inquiry and, consequently, in the long run, work to the disadvantage of society's general welfare."[7]

We have attempted to follow a scientific method in this work, and in particular we have directed our attention to the presuppositions and abstractions underlying the concepts of IQ and heritability. We have argued that these concepts systematically abstract from the real historical process in which humanity has developed through progressive transformation of nature and in social struggles for more advanced forms of society. In this overall social-historical process, the struggle for more advanced ideas and for the cultural development of the large majority of the people has played an essential role. Consequently, it is against the spirit of "free inquiry" to deprive individuals of the real "means of mental production" through increasing restrictions of access to institutions of higher education and through cut-backs in the quality of education in the public school systems. Real freedom of action as well as of thought require the availability of the objectively existing instruments of action and thought. There must be academic freedom and freedom of inquiry as much for Blacks as there is for whites. The genetic "meritocracy" theory is not only "undemocratic"—it runs counter to real freedom of inquiry, real academic freedom, real freedom of speech. It holds that such freedoms should be reserved for the few, while the majority are said to be incapable of benefiting from access to the means of more advanced intellectual development.

Such an argument runs counter not only to "ideals" and "egalitarian principles," but it contradicts the basic needs of society today. The development of the modern productive forces of today's society increasingly requires a more deeply and more broadly educated work force and poses immense challenges for the development of science. But the organization of society, based on the principle that the short-run interests of the few take precedence over the basic and long-run interests of the many, leads to the restriction of education and to the stifling of the sciences and culture.

Jensen accuses his opponents of not being "open" to the evidence. It is true that the abstractions that are built into the technical features of IQ and heritability theory make these concepts difficult to comprehend. But this is not because people are incapable of real thought. It is because they are not used to the truly *abstract* form of reasoning which is necessary to understand these concepts. It is because most people are not used to abstracting from real development, both the development of humanity and the development of the individual, in order to focus on theoretical formulae dealing with pure variations and inter-correlations of variations. This difficulty is evident in the numerous "common sense" criticisms of and outcries against Jensenism—some of which have been quoted in our text. These criticisms are not due to fastidiousness, intellectual taboos or desire for respectability. They are judgments based on real knowledge of real history. The impossibility of the genetic thesis stands out as a matter, not of dogma or ideology, but of historical reality. Based on such knowledge, the ordinary reasonable person finds the "genetic hypothesis," understood to mean what most people think it means, utterly unreasonable. Thus John Daniels and Vincent Houghton write: "If intellectual functioning is fixed at the moment of conception, if 80 per cent of adult performances is directly dependent upon genetic inheritance, how have the styles of our lives and the patterns of our thinking changed to the extent they have? Within psychometry, there is no answer."[8]

It is this kind of reasoning that finds the "genetic explanation" totally implausible from the beginning. The real difficulty for many is rather how some "experts" can propound theses that stand in sharp opposition to real historical evidence. It is important, certainly, to analyze the basic concepts which underlie "Jensenism"—which is only one form of several varieties of thought which today purport to give a biological interpretation of human society. If those outside of psychometrics have difficulty comprehending the abstractions from real history on which the "genetic explanation" is based, those who are "inside psychometrics" often have the opposite difficulty of seeing their concepts and measuring tools in a broader perspective. The technical difficulties with IQ and heritability abstractions create a labyrinth in which all too often a guiding thread is missing. The present book is not intended to oppose the measurement of human abilities, but to criticize metaphysical and idealistic theories of intelligence and accompanying educational practices. By putting this one approach to intelligence in proper perspective further progress can be made in the development of a real science of human intellectual processses. This in turn will be the basis for a more effective and a more humane system of education.

## *The practical reality of racism*

Ideas are not matters of purely theoretical interest. They have profound practical content and effects. In the relative isolation of the academic world, it is easy to focus on the formal aspects of theories and arguments and to overlook the reality which those ideas ultimately reflect. Intellectuals may be flattered by the notion that they are above obligation to the conditions and needs of society, and that their only obligation is to "pure science." But this illusion of the independence of the pursuit of knowledge only subjects institutions of intellectual activity all the more to bondage to the practical powers: the dominant class interests, which dispose of the ideas which scientists and academics propose. Commenting on the themes of his play, *The Life of Galileo,* Bertolt Brecht wrote: "The bourgeois single out science from the scientist's consciousness, setting it up as an island of independence so as to be able in practice to interweave it with *their* politics, *their* economics, *their* ideology. The research scientist's object is pure research; the product of that research is not so pure. The formula $E=mc^2$ is conceived of as eternal, not tied to anything. Hence other people can do the tying: Suddenly the city of Hiroshima became very short-lived. The scientists are claiming the irresponsibility of machines."[9]

When it is possible, as it was in the case of Brecht himself during the Cold War, this servitude is blatantly enforced on intellectuals

who are "open" to certain ideas—ideas that contradict the unre-
stricted rule of giant corporations over economic, political as well as
cultural life. If there is some greater liberty of thought in the colleges
and universities today, this is not because of a return to libertarian
values on the part of the corporate and political elite. It is the result
of the failure of the attempt of U.S. imperialism to "contain and roll
back" socialism in the 1950s—leading ultimately to the reluctant
and still inconsistent recognition of the necessity for the capitalist
world to peacefully coexist with socialism. It is the result of the
defeat of colonialism and the achievement of political independence
by the vast majority of the peoples of the world who had been
brutally subjected to imperialist exploitation. In the United States, it
follows from the courageous fight for elementary civil rights led by
Blacks against a vicious system of legal apartheid in the South. The
struggle for political and legal rights inevitably shifts to the fundamental struggle for full human rights—the rights of peoples to
control their own destinies by taking control of their resources, their
land and their industries. This movement to full social and economic
emancipation reached an historic turning point with the ending of
the war in Vietnam. With the moral and material assistance of the
socialist countries, the Vietnamese liberation movement dealt a
major blow to U.S. aims for controlling the riches of Southeast Asia.
Here students played an important part in the world-wide protest
against the injustice and barbarity of U.S. efforts to crush the
Indochinese movements for national and social liberation. But more
decisive was the fact that the "silent majority" of the U.S. population was becoming a vocal majority against a senseless and costly
war.

Each of these advances on the practical plane led to advances on
the plane of ideas, following intense ideological battles. Openness to
ideas that had been closed, that had been "taboo," became possible
as a result of historic efforts of millions of people. It had been taboo
to question the right of the United States to determine the destiny of
mankind. It had been taboo to oppose the idea that "national
security" justified any measures to protect mythical democracies
from an equally mythical "Communist aggression." The right of the
military-industrial complex to devour national resources as well as
lives has been a sacred cow, venerated by a powerful ideological
apparatus and backed up at certain times by the direct use of police

force. The powerful ideology of anti-Communism continues to prevent large numbers of people from being "open" to historic changes taking place in the world today as more and more peoples are taking the path to real independence and socialism.

As a means of disorienting the population, the pervasive force of anti-Communism is matched only by the equally powerful force of racism. Racism prevents white people from being open to their common humanity with people around the world. Racism inhibited the recognition that the "Asians" who were being covered with napalm in Indochina were also human beings with the same rights to self-determination that U.S. citizens expect for themselves. Racism prevents white people from seeing the special crimes being committed against the most exploited and oppressed populations in their own country, and blinds them to the fact that they are themselves an indirect victim of these crimes.

Racism creates a special kind of blindness—an inability to be "open" to certain realities. In the United States, the Black population earns scarcely more than half the average wages of whites, making possible enormous superprofits for the employers of Black labor as well as making it more difficult for white workers to defend and improve their own working and living conditions. With double the unemployment rate that exists for whites, Black people take the full brunt of economic crisis (together with other specially oppressed minority populations). As the economic crisis of the seventies deepens, the meagre advances on the economic and social planes won in past years are wiped out, and even legal advances are reversed. A system of dividing the population by race continues to exist and in recent years has even become worse.

Racism is too often seen as purely a matter of psychology and individual attitudes, while the practical discrimination in real life that breeds this psychology is overlooked. When Blacks fight against this system of special exploitation it is called discrimination in reverse. To see things in this light is an effect of the distorted vision caused by racism. Racism is both a socio-economic institution of U.S. history that has been perpetuated for four centuries, and the system of ideas that justifies this institution and renders people blind to its real nature. Racist ideology, which consists in blaming the victim, obscures the fact that racist practices are essentially a means of achieving superprofits from a major section of the population, and of dividing the working people against themselves so that all can be more thoroughly exploited. The fact that whites are also victims of racism is clearly seen in the U.S. South. There, racism has historically been most brutal and systematic. Far from benefiting from this, the southern white worker suffers from a lower standard of living than exists in the North, and has been deprived of basic labor rights. As more and more industries flee the northern cities for cheaper, unorganized labor regions of the South, the northern white worker too feels the effect of this aspect of racism. As we have seen, the ideology of IQ, which has a special impact on Blacks, is fundamentally directed against the working class as a whole.

Giving the white person a spurious sense of superiority over the Black, racism blinds him to his own common victimization as a fellow wage slave hired by the same boss. As economic contradictions inherent in capitalism cause both to suffer, but the Black still more than the white, the system of dividing the working population by race encourages the white worker to blame the Black for his ills. The fostering of racist ideas in both covert and more aggressive forms can become a powerful tool for preserving the class rule of the corporate monopolies when economic, political and social crises are calling this rule more and more into question. This is a dangerous weapon that turns one section of the working population on another, the scapegoat for its problems. And as the experience of Nazi racism shows, this is a weapon that can be used to promote the most brutal and self-destructive wars.

What kind of "openness" is it then which remains blind to these historic realities? If there is real openness, should it not be to the fact that racism has been and continues to be an integral fact of U.S. history, as it is endemic to capitalism as a whole? Marx wrote that the white worker can never be free while the Black worker is in chains. Nor can there be any real freedom of thought unless the facts of racism and the ideology of racism are exposed and destroyed. This means that there must be more than declarations regarding the equality of the races. There must be real social equality, which is only possible by taking special measures to counteract the continuing realities of racism.

Operation Headstart is one such measure. In the field of education, much more can be done, starting with the desegregation of the school system. Real affirmative action measures are necessary to overcome historic inequalities in hiring. On the plane of ideas, there is a pressing need for systematic education in the schools on the history, scope and real significance of racism. This is what real openness on the level of ideas means, if one has any consciousness of the facts of U.S. history and the fundamental problems of U.S. society. "Openness" to the idea that these facts and these problems stem from the innate characteristics of Black people themselves is worse than the narrowest closed-mindedness. It is an insult and slander directed against a proud people who have played a major role in shaping U.S. history, and in creating the values of democracy and liberty to the extent that these are more than hypocritical words. If the freedom that exists on the campuses is not to be suffocated by racism, as it is in South Africa and as it was in Hitler's Germany, it is essential that intellectuals make a searching examination of their responsibilities in promoting a culture free of the poison of racism.

## Too little intelligence or too much?

In his 1969 article, "How much can we boost IQ and Scholastic Achievement," Arthur Jensen speaks of the scarcity of intelligence in terms which only a few years later show what little historical understanding is contained in his analysis: "perusal of the want-ads in any metropolitan newspaper reveals that there are extremely few jobs advertised which are suitable to the level of education and skills typically found below IQs of 85 or 90, while we see everyday in the want-ads hundreds of jobs which call for a level of education and skills typically found among school graduates with IQs above 110. These jobs go begging to be filled. The fact is, there are not nearly enough minimally qualified persons to fill them."[10] There follows a discussion of the question as to whether the national intelligence is declining as a result of over-breeding by the "intellectually inferior." (After examining arguments which refute this theory, Jensen concludes that it may still be valid, especially in the Black population.) Today, with thousands of Ph.D.s looking for work or hired in unskilled or semi-skilled jobs, the perspective is clearly one of an *overproduction* of intellectual abilities and an incapacity on the part of a crisis-ridden economy to utilize the human potentialities which it has created. Already in 1969, the question of whether compensatory education programs "boosted IQ" was asked in the context of changing national priorities away from education—and especially from the minimal educational rights which Black Americans won during the civil rights movement. The Nixon administration was already preparing for the wage-freeze offensive, in conjunction with the building of an anti-democratic presidential power. It is not a matter of there being too little intelligence, but of justifying an effort to stifle the growth of highly qualified people, and of justifying anti-democratic trends in education which make it more and more difficult for those without the economic means to develop their intellectual abilities. And in a situation where there is scarcity not of *abilities* but of *places* in higher education for youth without substantial economic means, the insinuation that special programs or priorities that still exist for minority youth are being wasted on the intellectually inferior promotes racist division between Blacks and whites. The concept of the intellectual inferiority of Blacks encourages whites to fight Blacks for shrinking educational opportunities, rather than to defend the legitimate special demands of Blacks and to oppose the vicious system of racism. It is only by recognizing and acting upon the specific demands of Blacks that the kind of unity of Black and white can be developed which will be able to advance the interests of the majority of people.

Thus, the real issue underlying the discussion of "educability and group differences" is the surplus of intellectual abilities which cannot be absorbed by an economy that is stagnating and slipping into depression.

There are two main trends that emerge in this context. One which emphasizes hierarchy or "meritocracy"—with the implication that the few on top are there because of nature, because they were born with the brains, while the rest, clustering in a more and more mediocre middle or slipping to the bottom, have only themselves or their genes to blame. Another which promotes further advances in the democratization of education, responding to the real needs of development in modern society, and which begins to locate the obstacles to this development not in the people as a whole, but in a socio-economic system that subordinates the needs of the vast majority to the private interests of the few.

# Notes

## 1 • Jensen's claims

1. A. R. Jensen, "How Much Can We Boost I.Q. and Scholastic Achievement?" *Harvard Educational Review*, 39, 1969, p. 2. Also cf., Leon J. Kamin, *The Science and Politics of I.Q.*, John Wiley & Sons, New York, 1974. Collections of critical essays include N.J. Block & Gerald Dworkin, eds., *The IQ and Controversy*, Pantheon Books, New York, 1976; Gartner, Greer and Riessman, eds., *The New Assault on Equality*, Perennial Library, New York, 1974; Richardson and Spears, eds., *Race and Intelligence*, Penguin Books, Inc., Baltimore, Md., 1972; Tobach, Gianutsos, Topoff and Gross, *The Four Horsemen: Racism, Sexism, Militarism and Social Darwinism*, Behavioral Publications, New York, 1974. For a Marxist perspective, cf. Brian Simon, *Intelligence, Psychology and Education, A Marxist Critique*, Lawrence and Wishart, London, 1974. For a more general theoretical perspective which was extremely helpful to this author, cf. Lucien Sevé, *Marxisme et Theorie de la Personnalite*, Editions Sociales, Paris, 1969.

2. Ibid., p. 88.

3. A. R. Jensen, *Educability and Group Differences*, Harper & Row, New York, 1972, p. 6.

4. H.J. Eysenck, *The IQ Argument: Race, Intelligence and Education*, The Library Press, LaSalle, Illinois, 1971, p. 62.
   Sir Cyril Burt, the "father of British educational psychology," exercised incalculable influence on the testing movement from 1909 until his death in 1971. A teacher of Hans Eysenck, who in turn was a teacher of Arthur Jensen, Burt "wielded considerable influence over national educational policy. As a government advisor in the 1930's and 1940's, he was influential in setting up the three-tier system of British education. In accordance with Burt's views that intelligence is largely innate, children were irredeemably assigned to one of the three educational levels on the basis of a test given at the age of 11." (Nicholas Wade, "IQ and Heredity: Suspicion of Fraud Beclouds Classic Experiment," *Science*, Vol. 194:916-19, November 26, 1976.) These concepts and practices were challenged by democratic and Marxist educational theorists since the 1950s (cf. B. Simon, op. cit.). Burt's authority within mainstream academic circles remained unchallenged, however, until 1972 when Leon Kamin, a Princeton University psychologist, exposed distortions, falsifications and completely unscientific procedures in Burt's important studies of identical twins raised in separate environments, which is probably the main support for the hereditarian thesis. Beginning in October 1976, the London

*Sunday Times* printed a series of exposures, citing additional evidence of fraud. In particular, scarcely any trace can be found of the alleged co-authors of Burt's twin studies. "Should the accusation prove true," Wade writes, "the forgery may rank with that of the Piltdown Man in that for years it remained undetected while occupying a pivotal place in a fierce scientific controversy." For detailed discussion of Kamin's analysis, see chapter 10 of this book.

5. Cf. Berkeley Rice, "The High Cost of Thinking the Unthinkable," in *Psychology Today,* December 1973.
6. Jensen, *Educability and Group Differences,* op. cit., p. 7.
7. R.J. Herrnstein, *IQ in the Meritocracy,* Little, Brown and Co., Boston, 1973, p. 43.
8. Eysenck, *The IQ Argument,* op. cit., p. 55.
9. Cited by Eysenck, *The Inequality of Man,* Temple Smith, London, 1973, p. 11.
10. For a more balanced account of this period in the U.S.S.R. written by a non-Marxist, cf. L.R. Graham, *Science and Philosophy in the Soviet Union,* Knopf, New York, 1972.
11. E. Sharp, *The IQ Cult,* Coward, McCann & Geoghegan, New York, 1972.
12. O. Cohen, *Intelligence: What Is It?,* M. Evans, New York, 1974.
13. Cited in Simon, *Intelligence, Psychology and Education, A Marxist Critique,* op. cit., p. 261.
14. Jensen, *Educability and Group Differences,* op. cit., p. 6.
15. Joanna Ryan, "IQ—The Illusion of Objectivity," in *Race and Intelligence,* op. cit., p. 40.

## 2 • General theory and method of IQ

1. Eysenck, *The IQ Argument,* op. cit., p. 56.
2. Frederick Engels, *Ludwig Feuerbach and the Outcome of Classical German Philosophy,* International Publishers, New York, 1970, p. 45.
3. Ibid., p. 44.
4. Ibid., p. 25.
5. Jensen, "How Much Can We Boost I.Q. and Scholastic Achievement?", op. cit., pp. 5-6.
6. Ibid., p. 8; Jensen is citing favorably Edwin G. Boring.
7. Eysenck, *The Inequality of Man,* op. cit., p. 46.
8. Ibid., p. 47.
9. Cf. John Hoffman, *Marxism and the Theory of Praxis,* International Publishers, New York, 1975.
10. Jensen, "How Much Can We Boost I.Q. and Scholastic Achievement?", op. cit. p. 19.
11. Eysenck, *The IQ Argument,* op. cit., p. 25.
12. Ibid., pp. 49-50.
13. Frederick Engels, *Anti-Duhring,* International Publishers, New York, 1966, p. 106.
14. For introductory books on Marxist philosophy, cf. George Politzer, *Elementary Principles of Philosophy,* International Publishers, 1976; Maurice Cornforth, *The Theory of Knowledge,* International Publishers, 1971. For more advanced material in relation to currents in contemporary philosophy, cf. Hoffman, op. cit.; Cornforth, *The Open Philosophy and the Open Society: A Reply to Dr. Karl Popper's Refutations of Marxism,* International Publishers, New York, 1970; Cornforth, *Marxism and the Linguistic Philosophy,* International Publishers, New York, 1967; J. Lawler, *The Existentialist Marxism of Jean-Paul Sartre,* B.R. Grüner N.V., Nieuwe Herengracht 31, Amsterdam, 1976. U.S. distributor: Humanities Press, Atlantic Highlands, New Jersey.
15. Jensen, "How Much Can We Boost I.Q. and Scholastic Achievement?", op. cit., p. 6.
16. Engels, *Anti-Duhring,* op. cit., pp. 106-7.

## 3 • Some IQ questions

1. Quoted from Frank S. Freeman, *Theory and Practice of Psychological Testing*, Holt, Rinehart and Winston, Inc., New York, 1960, pp. 105-107.
2. The following material is contained in Terman and Merrill, *The Stanford-Binet Intelligence Scale: Manual for the Third Revision*, Houghton Mifflin Company, Boston, 1960.
3. Jerome Kagan, "The Magical Aura of IQ," in A. Montagu, ed., *Race and IQ*, Oxford University Press, 1975, pp. 54-55.
4. Ibid., p. 55.
5. Eysenck, *The IQ Argument*, op. cit., p. 50.
6. H. J. Eysenck, *Check Your Own IQ*, Pelican, New York, 1972.

## 4 • History of IQ theory

1. Herrnstein, op. cit., p. 63.
2. Ibid.
3. Ibid., p. 66.
4. Sharp, op. cit., p. 40.
5. Herrnstein, op. cit., p. 66
6. Sharp, op. cit., p. 41.
7. Cited by Kamin, op. cit., p. 6.
8. Ibid., p. 16.
9. The diagram is a simplified version of the bell curve which can be found in standard texts on statistics and educational psychology.
10. Cf. Allan J. Edwards, *Individual Mental Testing*, II, Intext Educational Publishers, Scranton, 1972, p. 5.
11. Lewis M. Terman, *The Measurement of Intelligence*, Houghton Mifflin Company, 1916, pp. 53-4.
12. Cf. Ross A. Evans, "Psychology's White Face," in *The New Assault on Equality*, op. cit.
13. "Tests involving the 'more complex mental processes' continued to give the best results in differentiating persons *judged by other criteria* to be more intelligent." Terman and Merrill, *The Stanford-Binet Intelligence Scale: Manual for the Third Revision*, op. cit., p. 6. (Emphasis added—J.L.)

## 5 • "Reliability" and "validity" of IQ tests

1. Philip E. Vernon, *The Measurement of Abilities*, University of London Press, Ltd., 1965, p. 148.
2. "Binet remarks that most people who inquire into his method of measuring intelligence do so expecting to find something very surprising and mysterious; and on seeing how much it resembles the methods which common sense employs in ordinary life, they heave a sigh of disappointment and say, 'Is that all?' Binet reminds us that the difference between the scientific and unscientific way of doing a thing is not necessarily a difference in the *nature* of the method; it is often merely a difference in *exactness*. Science does the thing better, because it does it more accurately." This early remark of Terman (*The Measurement of Intelligence*, op. cit., pp. 33-4) would have justified the ancient Ptolemaic astronomy, which specialized in "exact" descriptions of the "heavenly spheres" which were directly perceived in an "inexact" manner by "common sense."
3. Anne Anastasi, *Psychological Testing*, The Macmillan Company, New York, 1961, p. 135.
4. Ibid., p. 136.
5. Ibid., p. 137.
6. David C. McClelland, "Testing for Competence Rather Than for 'Intelligence,'" in *The New Assault on Equality*, op, cit.
7. Noting that IQ is "fairly stable under normal conditions of upbringing—more so than would be gathered from the publications of some left-wing theorists and of

some extremely environmentalistic American investigators," Vernon (op. cit., pp. 156-57) adds that there is, nevertheless, a significant fluctuation of scores. Between ages 6 to 11, and between ages 11 to 18, the average child fluctuates 7 IQ points up or down, while 17 percent fluctuate 15 or more points. It should be noted that by selecting the above age ranges Vernon minimizes the amount of fluctuation—which would be much higher between the ages of 6 to 18. It is also evident that such "stability" occurs "under normal conditions." It is precisely the thesis of "left-wing theorists" and environmentalists that significant changes would be dependent upon changing the "normal conditions."

8. For a comparative study of the educational methods in the USA and the U.S.S.R., which raises questions regarding such values in the U.S. system, cf. Urie Bronfenbrenner, *Two Worlds of Childhood, U.S.A. and U.S.S.R.*, Russell Sage Foundation, New York, 1970. The classic account of the early development of Soviet educational theory is A.S. Makarenko's captivating work, *The Road to Life (An Epic of Education)*, Progress Publishers, Moscow, 1973, in English.
9. Anastasi, *Psychological Testing*, op. cit., p. 144.
10. Ibid., p. 143.
11. Terman, *The Measurement of Intelligence*, op. cit., p.54. (Emphasis added—J.L.)
12. Anastasi, *Psychological Testing*, op. cit., pp. 145-46.
13. Jensen, "How Much Can We Boost I.Q. and Scholastic Achievement?" op. cit., p. 82.
14. Anastasi, *Psychological Testing*, op. cit., pp. 147-48.
15. Jensen, "How Much Can We Boost I.Q. and Scholastic Achievement?", op. cit., pp. 8,9.
16. Ibid., p. 9.
17. Ibid.
18. Anastasi, *Psychological Testing*, op. cit., p. 148.
19. Jensen, "How Much Can We Boost I.Q. and Scholastic Achievement?", op. cit., p. 3n.
20. Anastasi, *Psychological Testing*, op. cit., pp. 151-52.
21. Ibid., 1969 edition, p. 16.
22. Ibid., p. 23.
23. Anastasi, *Psychological Testing*, op. cit., p. 12.

## 6 • Differences: in the children or in the schools?

1. Jensen, *Educability and Group Differences*, op. cit., p. 74.
2. Jensen, "How Much Can We Boost I.Q. and Scholastic Achievement?", op. cit., p. 7.
3. Cf. ibid., p. 117.
4. Eysenck, *The Inequality of Man*, op. cit., p. 23.
5. Jensen, "How Much Can We Boost I.Q. and Scholastic Achievement?", op. cit., p. 10.
6. Ibid., p. 13.
7. Ibid.
8. Ibid.
9. Anastasi, *Psychological Testing*, op. cit., p. 262.
10. In R. Cancro, ed., *Intelligence: Genetic and Environmental Influences*, Grune & Stratton, New York, p. 269.
11. Anastasi, *Psychological Testing*, op. cit., p. 266.
12. A. R. Jensen, "The Differences are Real," in *Psychology Today*, December 1973, pp. 81-2.
13. Eysenck shows different brain wave patterns for high, average and low IQ children. He leaves it to the reader to draw the incorrect inference that such patterns preexisted the development of definite skills. In fact, the acquisition of definite skills *produces* neuro-physiological connections in the brain. (*The IQ Argument*, op. cit., p. 53)
14. George Thomson, *The First Philosophers*, International Publishers, New York, 1955, pp. 188-89.
15. Frederick Engels, *The Origin of the Family, Private Property and the State*, International Publishers, New York, 1972, p. 83.

16. Frederick Engels, *Dialectics of Nature,* International Publishers, New York, 1973, p. 19.
17. Ruth Beard, *An Outline of Piaget's Developmental Psychology,* Mentor, New York, 1972, p. 16.
18. Cf. Simon, *Intelligence, Psychology and Education, A Marxist Critique,* op. cit.
19. K. Marx and F. Engels, *Collected Works,* Vol. V, International Publishers, New York, 1976, p. 59.

## 7 • Dialectical relation of biology and society

1. *Psychology Today,* op. cit., p. 81.
2. Ibid., p. 86.
3. Jensen, *Educability and Group Differences,* op. cit., p. 28.
4. Engels, *Anti-Duhring,* op. cit., p. 79.
5. Cf. A. Montagu, "Introduction," *Race and IQ,* op. cit.
6. Karl Marx, *The Economic and Philosophical Manuscripts of 1844,* International Publishers, New York, 1972, p. 113.
7. Jensen, "How Much Can We Boost I.Q. and Scholastic Achievement?", op. cit., p. 43.
8. Cf. e.g., A. R. Luria, *The Role of Speech in the Regulation of Normal and Abnormal Behavior,* Liveright Publishing Corporation, 1961; Luria and Yudovich, *Speech and the Development of Mental Processes in the Child,* Staples Press, London, 1968; Cole, Maltzman, eds., *A Handbook of Contemporary Soviet Psychology,* Basic Books, Inc., New York, 1969.
9. Marx, *Manuscripts,* op. cit., pp. 140-41.
10. Ibid., p. 141.
11. Engels, *Origin of the Family,* op. cit., p. 117.
12. M. Nesturkh, *The Races of Mankind,* Foreign Languages Publishing House, Moscow, pp. 57-8.
13. Ibid., p. 58.
14. Marx, *Manuscripts,* op. cit., p. 111.
15. Ibid., p. 122.
16. Cf. Victor Perlo, *Economics of Racism U.S.A.,* International Publishers, New York, 1975.
17. Ibid., p. 142.
18. Engels, *Dialectics of Nature,* op. cit., p. 19.

## 8 • Relative and absolute differences

1. Herrnstein, op. cit., p. 74.
2. Ibid., p. 90.
3. *Hegel's Science of Logic,* Humanities Press, 1969, pp. 218-9.
4. Herrnstein, op. cit., pp. 67-8.
5. B. M. Teplov and V. D. Nebylitsyn, "Investigation of the Properties of the Nervous System as an Approach to the Study of Individual Psychological Differences," in *Handbook of Contemporary Soviet Psychology,* op. cit.
6. Jensen, "How Much Can We Boost I.Q. and Scholastic Achievement?", op. cit., p. 89.
7. Engels, *Dialectics of Nature,* op. cit., pp. 184-85.
8. K. Marx and F. Engels, *Collected Works,* Vol. V, op. cit., p. 394.
9. Herrnstein, op. cit., p. 74.
10. Karl Marx, *A Contribution to the Critique of Political Economy,* International Publishers, New York, 1970, pp. 205-6.
11. Jensen, "How Much Can We Boost I.Q. and Scholastic Achievement?", op. cit., p. 17.
12. Ibid., p. 21.
13. Ibid., pp. 19-20.
14. Ibid., p. 88.

15. Ibid.
16. Engels, *Dialectics of Nature*, op. cit., p. 185.
17. Jensen, "How Much Can We Boost I.Q. and Scholastic Achievement?", op. cit., p. 22.
18. Cf. Eysenck, *The Inequality of Man*, op. cit., p. 223.
19. Ibid., p. 9.
20. Cf. J. Lawler, "Dialectical Philosophy and Developmental Psychology," in *Human Development*, Vol. 18, No. 1-2, 1975.
21. Engels, *Anti-Duhring*, op. cit., p. 19.
22. Karl Marx, *Capital*, Vol. 1, International Publishers, New York, 1967, p. 79.
23. "Where the political state has attained its true development, man—not only in thought, in consciousness, but *in reality*, in *life*—leads a twofold life, a heavenly and an earthly life: life in the *political community*, in which he considers himself a *communal being*, and life in *civil society*, in which he acts as a *private individual*, regards other men as a means, degrades himself into a means, and becomes the plaything of alien powers." Karl Marx, "On the Jewish Question," cited in *Collected Works*, Vol. III, International Publishers, New York, 1975, p. 154.
24. The following discussion is the author's interpretation of the relevant sections of *Capital*, op. cit., in three volumes.
25. Engels, *Dialectics of Nature*, op. cit., p. 19.
26. Engels, *Ludwig Feuerbach*, op. cit., p. 49.

## 9 • Heritability

1. Cf. Claude Lightfoot, *Racism and Human Survival*, International Publishers, New York, 1972. While reviewing the history of racism in Nazi Germany, and providing a useful bibliography of important works criticizing racist theories, this book focuses primarily on the contrasting approaches to the problem of racism in the two Germanys today; where active anti-racist education characterizes the educational curriculum in the socialist German Democratic Republic, "benign neglect," to say the least, is characteristic of the failure of the Federal Republic of Germany to come to grips with the continuing presence of racism.
2. T. Dobzhansky, *Genetic Diversity and Human Equality*, Basic Books, New York, 1973, p. 21.
3. Jensen, *Educability and Group Differences*, op. cit., p. 3.
4. The same results occur when genetically identical corn is raised in different environments successively. The different performances of the same genotype under different environmental conditions (say, by the gradual increase in the amount of nitrogen) is called the "norm of reaction." Jensen assumes that for human beings these are linear—that as one enriches the environment, an individual with a superior genotype will develop faster than an individual with an inferior genotype (for IQ). This notion has been criticized as a pure assumption, and quite atypical for genotypes of the non-intellectual type. In any case a norm of reaction measures already given environmental conditions and cannot predict for as yet untested environments. It is an essentially descriptive concept.
5. Richard Lewontin, "Race and Intelligence," in *The IQ Controversy*, op. cit.; Jensen's reply and a rejoinder by Lewontin are contained in the same volume.
6. Cited by Simon, op. cit., p. 258.
7. In Engels, *Ludwig Feuerbach*, op. cit., p. 83.
8. Howard Topoff, "Genes, Intelligence and Race," in *The Four Horsemen*, op. cit.
9. Jensen, *Educability and Group Differences*, op. cit., p. 44.
10. Ibid., p. 51.
11. W. F. Overton, "On the Assumptive Base of the Nature-Nurture Controversy: Additive versus Interactive Conceptions," *Human Development*, 16: 74-89, 1973. The Anastasi article is from *Psychological Review*, 65: 197-208, 1958.
12. Jensen, *Educability and Group Differences*, op. cit., p. 52.
13. Ibid., p. 58.
14. Ibid.

15. Ibid., pp. 49-50.
16. Topoff, op. cit., p. 27.
17. Ibid., p. 42.
18. Cf. eg., David Paterson, *Applied Genetics,* Doubleday and Co., Inc., New York, 1969. Paterson expresses great caution with respect to the application of genetics to human life—with the exception of genetically based diseases and medical applications, and is basically critical of a eugenic approach to human life. Other popular literature, however, puts the future of mankind in the hands of the geneticists. Cf. D. S. Halacy, Jr., *The Genetic Revolution, Shaping Life for Tomorrow,* Harper & Row, New York, 1974. Halacy sees the key to man's control of his life in genetics and attributes the reserve of many scientists to a conspiracy of silence. Jensen et al are presented as victims of dogmatism. We should not let the association of eugenics with Nazism and racism prevent us from forging ahead with genetic engineering, argues Halacy.
19. John Hambley, "Diversity: A Developmental Perspective," in *Race and Intelligence,* op. cit. Hambley describes the relation between genotype and phenotype as "indeterminate." While critical of the simplicity of the idea of the fixed heritability of IQ, Hambley remains within a biological framework by urging society to maximize its genetic diversity.
20. K. Marx and F. Engels, *Collected Works,* Vol. V op. cit., p. 3.

## 10 • Twins and other relations

1. Jensen, "How Much Can We Boost I.Q. and Scholastic Achievement?", op. cit., p. 49.
2. Cf. Herrnstein, *IQ in the Meritocracy,* op. cit., pp. 82-3.
3. Jensen's response to the argument that within-group-heritability cannot be used to estimate the cause of differences between groups is that it is a merely *formal* objection, and while nothing can be *proven* by heritability figures for within group variations they at least increase the *plausibility* that the cause of group differences is heritable. This response, clearly refuted by Lewontin's examples, shows how Jensen is trying to load a technical, formal and relative definition with his own metaphysical and speculative interpretations. Cf. Jensen, *Educability and Group Differences,* op. cit., pp. 133-48.
4. Cf. Engels, *Origin of the Family,* op. cit.
5. On a more general level this idea was developed by Hegel in his opposition to phrenology and other biological or physiological explanations of human behavior. Hegel argued that "natural disposition" played a role in explaining behavior to the extent that the relevant activities involved the least developed historical content. In this category Hegel included mathematics and music, where, he noted, child prodigies were most often found. Cf. volume 3 of Hegel's *Encyclopedia, The Philosophy of Mind.*
6. Cf. Kamin, op. cit., "I.Q. in the Uterus," pp. 161-74.
7. In a paper delivered by invitation to the American Psychological Association, in Washington, D.C., on September 4, 1976, Robert C. Nichols compares scores on the National Merit Qualifying Test of identical and fraternal twins of the same sex. Arguing that there are no significant environmental differences between identicals and fraternals, Nichols concludes from the fact that the scores of the former correlate more closely than those of the latter that heritability is substantial (.50). In addition to the inadequacy of the main premise of the paper—that there are no consistent environmental differences between identical and fraternal twins— Nichols' paper to the APA asserts that the twins studied are "not representative of any specific group to which statistical generalizations might be made" (p. 8). This claim contradicts the assertion of class bias due to the largely middle-class origins of the test takers which is admitted in the book from which the article is derived: Loehlin and Nichols, *Heredity, Environment and Personality,* University of Texas Press, Austin and London, 1976. The point is this: even if there were heritability *within* a given social class this would not prove that the differences between

classes were heritable—unless one could assume that environmental differences were of the same kind and could be estimated quantitatively. But this last assumption denies qualitative differences between classes. Suppose that some children are more "nervous" than others as a result of genetically influenced production of greater amounts of a chemical whose effects on the nervous system are similar to those of caffein. Such effects are of course relative to diet and upbringing which not only vary quantitatively between social classes, but could produce qualitatively different and even opposite results. Thus, the "nervous" child from the middle-class family with special educational advantages may channel additional energy in a scholastically productive manner. The opposite could conceivably occur for the working-class child.

An unpublished paper, "Twins and Survival: A Response to Professor Robert C. Nichols on the Subject," is available from this author.

8. Jensen, "How Much Can We Boost I.Q. and Scholastic Achievement?", op. cit., pp. 47, 52.
9. Kamin, op. cit., p. 44.
10. Ibid., p. 42.
11. Ibid., p. 67.
Responding to accusations of fraud against Burt by a team of investigators for the London *Sunday Times,* led by a geneticist and medical correspondent, Dr. Oliver Gillie, Jensen has come to Burt's defense. In a letter to the *The London Times,* Jensen claims that he himself published all the "errors or inconsistencies" in a 1974 article, neglecting to mention here that it was Kamin who prompted this interest. (Jensen acknowledges this in a footnote to the 1974 article.) Jensen attributes methodological failings to the carelessness of an old man. He argues that the elimination of Burt's work from psychological literature would have no effect on findings from other data. This argument neglects 1) the enormous authority and power exercised by Burt for sixty years on the general orientation of psychometrics; 2) the fact that only Burt's 1966 paper alleges that the environments of his 53 twins was totally uncorrelated—an absolutely essential requirement for any assertions regarding the heritability of a population. Cf. Wade, op. cit., and Oliver Gillie and Arthur R. Jensen, "Did Sir Cyril Burt Fake His Research on Heritability of Intelligence?", *Education Digest,* Vol. 42: 43-5, March 1977.
12. The combined score, used by Jensen, is listed as .77, obscuring the clear environmental effects revealed when the two categories of "separation" are distinguished.
13. Kamin, op. cit., p. 50.
14. Ibid., p. 53.
15. Ibid., pp. 61-2.
16. Ibid., p. 75.
17. Ibid., p. 78.
18. Ibid.
19. Ibid., p. 79.
20. Ibid., p. 134; cf. Herrnstein, op. cit., p. 150.

## 11 • Real science and real freedom

1. Jensen, *Educability and Group Differences,* op. cit., p. 4.
2. Ibid.
3. Ibid.
4. Ibid., pp. 4-5.
5. Ibid., p. 5.
6. Ibid., p. 15.
7. Ibid.
8. John Daniels and Vincent Houghton, "Jensen, Eysenck and the Eclipse of the Galton Paradigm," in *Race and Intelligence,* op. cit., p. 74.
9. Bertolt Brecht, *Collected Plays,* Vol. 5, edited by Ralph Manheim and John Willett, Pantheon Books, New York, 1972, p. 220.
10. Jensen, "How Much Can We Boost I.Q. and Scholastic Achievement?", op. cit., p. 88.

# Index

## A

Ampere, Andre-Marie, pg. 22
Anastasi, Ann, pp. 56, 57, 59, 60, 61, 62, 63, 65, 66, 67, 71, 74, 146
Aristotle, pp. 76, 81, 128

## B

Beard, Ruth, pg. 79
Binet, Alfred, pp. 29, 40, 42, 43, 44, 45, 46, 47, 48, 49, 51, 107, 108, 109
Brecht, Bertolt, pg. 176
Brown, Roy, pg. 141
Burt, Cyril, pp. 8, 165, 166, 169

## C

Cattell, Raymond B., pp. 70, 71
Copernicus, pg. 9

## D

Daniels, John, pg. 175
Darwin, Charles, pp. 9, 39, 77, 89, 90, 91, 102, 129, 152
Dobzhensky, T., pg. 137

## E

Einstein, Albert, pg. 9
Engels, Frederick, pp. 9, 18, 19, 24, 25, 27, 77, 78, 90, 91, 92, 96, 97, 101, 109, 113, 123, 125, 129, 132, 152, 162
Eysenck, Hans Jurgen, pp. 8, 9, 11, 12, 17, 21, 23, 24, 25, 35, 36, 42, 52, 70, 72, 73, 74, 79, 88, 121, 122, 168

## F

Faraday, Michael, pg. 22
Feuerback, Ludwig, pg. 157

## G

Galparin, Charles, pg. 79
Galton, Francis, pp. 39, 40, 41, 42, 43, 44, 45, 46, 47, 51, 69, 89, 133
Goddard, Henry, pg. 47
Goodenough, F. L., pg. 72

## H

Hambley, John, pg. 157
Hegel, Georg Wilhelm Friedrich, pp. 18, 106, 107, 123, 125
Herrnstein, Richard J., pp. 8, 9, 10, 23, 39, 40, 43, 44, 52, 79, 104, 106, 107, 108, 109, 115, 116, 121, 166, 168, 172
Hitler, Adolf, pg. 180
Houghton, Vincent, pg. 175

## J

Jarvik, Lissy F., pp. 168, 169
Jensen, Arthur R., pp. 7, 8, 9, 11, 12, 19, 20, 21, 23, 24, 26, 27, 36, 37, 52, 61, 62, 63, 64, 65, 67, 68, 69, 70, 71, 73, 74, 83, 84, 85, 87, 88, 89, 90, 91, 94, 104, 108, 112, 116, 117, 118, 119, 120, 121, 125, 133, 134, 135, 137, 139, 140, 141, 145, 146, 147, 149, 150, 151, 152, 157, 159, 160, 164, 165, 166, 168, 169, 173, 174, 175, 180
Jefferson, Thomas, pg. 9

**191**

**K**

Kagan, Jerome, pp. 33, 34
Kamin, Leon J., pp. 165, 166, 168, 169, 170, 171, 172

**L**

Leibniz, Gottfried W., pp. 112, 113
Lewontin, Richard, pp. 139, 142

**M**

Marx, Karl, pp. 9, 77, 82, 83, 93, 94, 95, 99, 100, 101, 114, 116, 127, 128, 131, 132, 142, 157, 158, 179
Maxwell, James C., pg. 22
McClelland, David, pp. 56, 57
Mendel, Gregor, pp. 89, 90, 151, 152, 153, 155
Morgan, Lewis Henry, pg. 77
Mozart, Wolfgang, Amadeus, pg. 94

**N**

Nebylitsyn, V.D., pg. 108
Nesturkh, M., pg. 97
Newton, Isaac, pp. 112, 113, 114
Nixon, Richard M., pg. 181

**O**

Ohm, Georg Simon, pg. 22
Overton, Willis F., pg. 146

**P**

Plato, pp. 76, 85

**R**

Ricardo, David, pg. 128
Rousseau, Jean Jacques, pg. 9

**S**

Simon, Brian, pp. 79, 80
Sims, Verner, pp. 169, 170
Smith, Adam, pg. 128
Spearman, Charles, pp. 62, 63, 117

**T**

Teplov, B.M., pg. 108
Terman, Lewis, pp. 46, 47, 48, 49, 50, 51, 60, 133
Thomson, George, pp. 75, 81
Thorndike, Edward Lee, pg. 89
Topoff, Howard, pp. 142, 151, 152